BASIC
PHYSICAL CHEMISTRY
The Route to Understanding

Revised Edition

T0322008

BASIC
PHYSICAL CHEMISTRY

The Route to Understanding

Revised Edition

E Brian Smith

Formerly Master of St Catherine's College, Oxford, UK &
Vice-Chancellor of Cardiff University, UK

Imperial College Press

Published by

Imperial College Press
57 Shelton Street
Covent Garden
London WC2H 9HE

Distributed by

World Scientific Publishing Co. Pte. Ltd.
5 Toh Tuck Link, Singapore 596224
USA office: 27 Warren Street, Suite 401-402, Hackensack, NJ 07601
UK office: 57 Shelton Street, Covent Garden, London WC2H 9HE

British Library Cataloguing-in-Publication Data
A catalogue record for this book is available from the British Library.

BASIC PHYSICAL CHEMISTRY
The Route to Understanding
(Revised Edition)

ISBN 978-1-78326-293-9
ISBN 978-1-78326-294-6 (pbk)

Typeset by Stallion Press
Email: enquiries@stallionpress.com

Printed in Singapore

For Regina

Preface to Revised Edition

This revised edition has enabled improvements and corrections to be made. There no doubt will be further changes and improvements to be identified and I would be grateful to be informed of these. Thank you in advance for e-mailing your comments to: e_brian_smith@ talktalk.net.

I would like to thank those who have given feedback. In particular I am greatly indebted to Dr David Gough for his invaluable help in preparing this edition, and to Jacqueline Downs for her editorial support.

E. Brian Smith

Preface to the First Edition

Physical chemistry textbooks for university students are typically substantial volumes, often 1000 pages long and in large format. That is the equivalent of nearly three quarters of a million words. Although, fortunately, some of that space is devoted to tables and diagrams, digesting material on such a scale nevertheless represents a formidable challenge for someone wishing to become familiar with the subject.

I have long planned to produce a book which would extend to the whole of physical chemistry the approach I used in *Basic Chemical Thermodynamics*. I set out my motivation in the preface to that book:

> The first time I heard about chemical thermodynamics was when a second-year undergraduate brought me the news early in my freshman year. He told a spine-chilling story of endless lectures with almost three-hundred numbered equations, all of which, it appeared, had to be committed to memory and reproduced in exactly the same form in subsequent examinations. Not only did these equations contain all the normal algebraic symbols, but, in addition, they were liberally sprinkled with stars, daggers, and circles so as to stretch even the most powerful of minds. Few would wish to deny the mind-improving and indeed character-building qualities of such a subject! However, many young chemists have more urgent pressures on their time.

This present book aims to introduce students to the principles that underpin physical chemistry in a similarly concise manner. Although based on current university courses in introductory physical chemistry I have aimed to give the book a distinct character. Fundamental principles are discussed in some detail and the book provides a solid platform from which to embark on further study of physical chemistry. Clearly, in the space of less than one third of the size of a conventional textbook, not all aspects of the subject can be treated fully. The book provides only brief accounts of the historical development of the subject and it does not include descriptions of the experimental observations on which our knowledge rests. The book attempts to steer the narrow course between bland superficiality and over-detailed notes. On occasion, it may encroach on one of these hazards or the other. For example, in dealing with molecular spectra, attention to detail seems necessary to do justice to the many applications of this subject in physical chemistry. Though some chapters draw on the material in *Basic Chemical Thermodynamics*, a different approach to introducing thermodynamics has been employed here, with greater emphasis on statistical and molecular aspects.

Every attempt has been made to avoid unnecessary mathematical manipulation. In order to emphasise this point, individual equations have not been numbered. When a reference to the origin of an equation is necessary, the section of the book in which it is introduced is given. A degree of duplication has been included to reduce the need for back reference. A working knowledge of only the most simple differential and integral calculus is required.

The nomenclature used in this book follows closely the recommendations of IUPAC, with a few minor exceptions. For example, I have followed the more traditional path of calling Gibbs energy 'Gibbs free energy' and have employed the label amu for atomic mass units. In addition, the superscript 0 has been used to identify the standard state of a pure substance at a pressure of one bar. Much of the older reference data is available at the standard state of one atmosphere. For the purposes of this book, 1 bar can be regarded as equivalent to 1 standard atmosphere, the difference of some 1% rarely being of significance.

I am grateful to Professor Sir John Rowlinson FRS, Professor Geoffrey Maitland and Dr Grant Ritchie for their comments on early drafts of the book. I also wish to thank Professor R. N. Zare for much inspiration when, some time ago, we developed the themes of another book, yet to be written. I would like to thank Mr Marco Piccinini for the great care he took in preparing many of the diagrams. I am particularly indebted to Mr Aleks Reinhardt who cast his eagle eye over the final manuscript, identifying many misprints, mistakes and inelegant arguments.

I am most grateful to my wife Regina, for her invaluable advice and for the unceasing support she has shown while this book was being written. The writer of popular children's books, Enid Blyton, describes a character called Uncle Quentin. Uncle Quentin is a scientist who does 'important work' and only leaves his study to demand that other occupants of the house keep quiet. I hope that, now that this book is completed, the ghost of Uncle Quentin will be banished from our home for some time to come.

In the introductory chapter, I compare physical chemistry to a plateau above challenging cliffs. I hope this book will help provide a guide to the routes through these intellectual barriers and lead students to a higher understanding of the subject.

E. Brian Smith
Oxford, 2011

Contents

Notation

The notation used in this book follows, with a few minor exceptions, IUPAC recommendations.

Symbols

Symbols used most extensively in the text are given below. Where, on occasion, the symbols have been used to represent different quantities, this is made clear in the text.

a activity

A Helmholtz (free) energy, atomic mass number

B magnetic field strength

C heat capacity

c concentration, root mean molecular speed

E energy, electromotive force

G Gibbs (free) energy

H enthalpy

$\mathbf{H}$ Hamiltonian operator

J, j total angular momentum

K equilibrium constant

L, l orbital angular momentum

M relative molar mass, 'molecular weight'

m mass, molality

n amount of substance

P pressure

p momentum, probability

Q reaction quotient, electric charge

q heat, electric charge

R, r distance

S entropy

T temperature

U internal energy

u intermolecular potential energy

V volume, potential energy

v velocity

w work

x mole fraction

Z nuclear charge, 'atomic number', collision rate, canonical partition function

z molecular partition function

γ activity coefficient

ε permittivity

θ quadrupole moment

λ wavelength

μ chemical potential, dipole moment, reduced mass

ν frequency

ξ extent of reaction

σ molecular collision diameter

Φ electrostatic potential

ψ wave function

States of matter

The state of matter is indicated by symbols in parentheses following the symbol for the relevant property.

gas   (g)
liquid   (l)
solid   (s)
solution   (soln)
aqueous solution   (aq)

Process

Thermodynamic processes are indicated by a subscript.

vaporization, boiling   vap
fusion, melting   fus
transition   trans
formation from elements   f
reaction   r

Thus, the enthalpy change on melting of a substance would be written $\Delta_{fus}H$. (The older notation, ΔH_{fus}, is still widely used).

Standard states

Three symbols are used as superscripts to identify standard states:

0   substance in a pure state at one bar pressure
*   substance in a pure state at an arbitrary pressure
⊖   arbitrary standard state (often a hypothetical state).

The most frequently used in this text are unit molality ($mol\,kg^{-1}$) and unit molarity ($mol\,L^{-1} \equiv mol\,dm^{-3}$).

The thermodynamic properties of elements and compounds are most commonly reported for a standard state of one bar pressure at 298.15 K. Much earlier data was provided at a standard state of one standard atmosphere pressure (1 atm). For most purposes, we can regard one bar and one atmosphere as equivalent (1 atm = 1.013 bar) and using this data does not, normally, introduce significant errors into thermodynamic calculations.

Components

Components of a system are indicated by subscripts. The subscript i is used to denote an unspecified chemical compound in equations of general applicability.

Units

Commonly-employed units and conversion factors are given in Appendix 4. The manipulation of units is greatly aided by what is called *quantity calculus*. Each quantity, for instance 5 m, is regarded as being composed of two elements: the number, five, and the unit, metres. When multiplying and dividing quantities, both of these elements follow the normal rules of algebra. For example, to evaluate the energy corresponding to a frequency of $5 \times 10^{14} \, \text{s}^{-1}$ using the relation $\Delta E = h\nu$, we can write

$$\Delta E = (6.6 \times 10^{-34} \, \text{J s})(5 \times 10^{14} \, \text{s}^{-1}) = 6.6 \times 10^{-34} \times 5 \times 10^{14} \, \text{J s} \times \text{s}^{-1} = 33 \times 10^{-20} \, \text{J}.$$

Quantity calculus makes the conversion from one unit to another quite straightforward. To convert a length l of 5 feet to metres, we write

$$l \, \text{m} = l \, \text{ft} \times (\text{m/ft}).$$

The conversion factor from feet to metres, (m/ft), is 0.30 m/ft and we obtain

$$l = 5 \, \text{ft} \times 0.30 \, (\text{m/ft}) = 1.5 \, \text{m}.$$

Conversion factors of use to chemists are given in Appendix 4.

When dealing only with quantities specified in SI units, we often do not need to specify the units of each quantity. The final result obtained by multiplying and dividing a number of quantities will appear in the appropriate SI units, determined by its dimensions.

Acknowledgements and Credits

A number of the figures are reproduced with the kind permission of various authors and publishers.

From: M. Rigby, E. B. Smith, W. A. Wakeham and G. C. Maitland, *Forces Between Molecules* (1986). Clarendon Press, Oxford.
Figures 2.7, 9.7, 9.9, 9.11 and 9.13.

From: C. N. Banwell, *Fundamentals of Molecular Spectroscopy*, 3rd Edition (1972). McGraw–Hill, London.
Figures 6.6, 6.12 and 6.15.

From: B. P. Straughan and S. Walker, *Spectroscopy*, Vol. 2 (1976). Chapman and Hall, London.
Figure 6.13.

From: D. A. McQuarrie and J. D. Simon, *Physical Chemistry: A Molecular Approach* (1997). University Science Books, Sausalito, California.
Figures 6.20, 9.14 and 11.13.

From: M. J. Pilling, *Reaction Kinetics* (1975). Clarendon Press, Oxford.
Figures 11.11 and 11.12.

From: P. J. Wheatley, *The Determination of Molecular Structure* (1981). Dover, New York.
Figure 9.17.

From: B. J. Alder and T. Wainwright, *Studies in Molecular Dynamics. I. General Method* (1958). Journal of Chemical Physics **31**:459, American Institute of Physics, New York.
Figure 9.4.

From: K. T. Gillen, A. M. Rulis and R .B. Bernstein, *Molecular Beam Study of the K + I₂ Reaction: Differential Cross Section and Energy Dependence* (1971). Journal of Chemical Physics **54**:2831, American Institute of Physics, New York. Figure 11.13.

From: G.-J. Su, *Modified Law of Corresponding States for Real Gases* (1946). Industrial and Engineering Chemistry **38**:803, American Chemical Society, Washington. Figure 9.12.

1

Background

1.1 Introduction

The boundaries of scientific subjects are defined only by usage. They change with time and so it is rarely worthwhile to seek detailed definitions. However, it can be said that chemistry is the study of the structure, composition and transformations of substances and we may take this as a useful working definition that will serve our purpose. Physical chemistry seeks to provide a quantitative understanding of chemistry. To do this requires an understanding of the structure of matter, of how atoms combine to form molecules and, when chemical compounds react, what determines the speed of reaction and what new substances result. We need to answer questions such as:

(i) why are there approximately one hundred elements each with different properties but with clear family relationships?

(ii) how are stable chemical compounds formed and what factors contribute to their stability?

(iii) why, when substances are exposed to light (or other radiation), are only some wavelengths absorbed, which may or may not induce chemical change?

(iv) why do some compounds react rapidly and completely to produce new substances and others do not react at all?

(v) why some compounds which can, in principle, react to produce a new stable product, do not do so, or at least react only very slowly?

(vi) why can substances exist in the gas, liquid or solid states?

It is possible to approach the subject of physical chemistry by reviewing the experimental evidence which provides the facts on which chemical theories must be based. An alternative approach starts with the theories that have been developed and seeks to interpret chemical phenomena on the basis of these theories. Given ample time and space, there is much to be said for the first approach because it corresponds to the way in which our understanding of chemistry has developed. However, given very limited space and time, it is much more economical to take the second approach

1

and identify the fundamental principles that underlie chemical theory and show how they can provide an excellent interpretation of the relevant phenomena.

Physical chemistry is a discipline that is perceived very differently by different people. Those who meet it when studying for degrees in biological subjects, or when undertaking medical training, often find it challenging, if not confusing, whereas those for whom it is a primary study find it elegant and coherent. One of the reasons for these quite different perceptions is clear. Most subjects, as we learn them, start off easy and become harder as we delve deeper. They are like walking up rugged mountains; easy at the bottom, but steeper and more difficult the higher we climb. Physical chemistry is rather different and it has been compared to a high plateau; most of its upper reaches are open and relatively easy to traverse. However, to reach these pastures, we must climb through formidable cliffs by following a steep and tortuous path (Fig. 1.1). Physical chemistry is founded on two powerful principles (and a few less important ones). When mastered, these principles provide the key to clarity and understanding. However, the principles involve inherently difficult concepts and both focus on uncertainty and probability, rather than offering certainty.

The first principle is based on the concept that *energy is not continuous* and can have only certain fixed values. It is said to be *quantised*. The consequences of this simple fact are widespread and quite dramatic in determining the behaviour

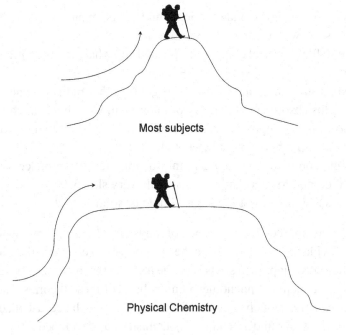

Most subjects

Physical Chemistry

Fig. 1.1. The path to understanding physical chemistry.

of chemical substances. The principle can also be stated in a seemingly quite different and more obscure way: '*Matter can exhibit both wave-like and particle-like behaviour.*' We do not see this in the world of large objects that surround us in everyday life, but it is important for small particles confined to a limited region of space. Elucidating the consequences of this statement lies at the foundation of quantum mechanics and will take us to the heart of physical chemistry.

The second principle is based on the fact that the behaviour of atoms and molecules in bulk is determined by statistical considerations. Thus, whereas there is no certainty that all the air molecules in a room will not congregate in one corner, we can be confident that this is very unlikely to happen. However, there is no rule governing the motion of molecules, which tells us that the air cannot go into a corner and leave someone in the centre of the room out of breath! The reason this does not happen is that it is less probable than when the air is more uniformly spread. The number of molecules is so large that the probability that this will not happen is overwhelming. In fact, this probability is so overwhelming that it effectively becomes a scientific certainty; an aspect of the Second Law of Thermodynamics.

To discover how these two principles lead to an understanding of the subject is a long journey and, to start, we must first describe the elements of basic chemistry and physics that will underpin our study. (Those familiar with elementary chemistry and physics might wish to skip much of this chapter.)

1.2 The building blocks

As chemists, we are fortunate that in a universe composed of so many particles characterised by different mass, charge, spin and other, more obscure, properties, we are able to explain most phenomena of importance in chemistry in terms of only four entities: the *electron*, the *proton*, the *neutron* and the *photon* (Table 1.1). The electron, proton and neutron are called *fermions*. These are particles which have half-integer values of spin. The photon is a *boson*, defined as a particle with zero or whole integer values of the spin. The angular momentum of a particle with a spin of 1/2 is $(3/4)^{1/2}\hbar$, where $\hbar = h/2\pi$. For a photon which has a spin of unity, the angular momentum is $(2)^{1/2}\hbar$. Atoms, though composed of fermions, may be either fermions or bosons. Fermions have the characteristic that they do not like being herded together; this is a property we will consider more rigorously later. This means that they tend to fill space and constitute what we identify as matter. On the other hand, bosons do not have this limitation and are the particles that transmit the forces which act between fermions. An analogy that is sometimes used is that of two skaters hitting tennis balls back and forth. The skaters would drift further apart

Table 1.1.   The fundamental particles.

Particle	Charge	Spin	Approx mass/amu	Exact mass/amu	Exact mass $\times 10^{27}$/kg	Class
proton	+1	1/2	1	1.00728	1.6726	fermion
neutron	0	1/2	1	1.00867	1.6749	fermion
electron	−1	1/2	0	0.000549	0.0009109	fermion
photon	0	1	0	0	0	boson

atomic mass unit $= 1.6605 \times 10^{-27}$ kg
elementary charge $e = 1.6022 \times 10^{-19}$ C
Approximate masses are used to calculate the mass numbers of atoms.
The unit of spin angular momentum is $h/2\pi$, where h is Planck's constant, 6.626×10^{-34} Js (see Section 3.2).

and an external observer would deduce a repulsive force acting between the skaters. The skaters represent fermions and the tennis balls, bosons.

There are four fundamental forces in nature: *gravity,* the *strong nuclear force,* which holds together the atomic nuclei, the *weak nuclear force,* which influences radioactive decay, and *electromagnetic* forces. (The last two forces can be shown to be closely related.) Despite its importance in cosmology, gravity is a weak force on the molecular scale and plays a negligible role in chemistry. The nuclear forces are short-ranged and their effects are essentially confined to within the nucleus. In chemistry, we are mainly concerned with the effects of electromagnetic forces which are transmitted by photons. We are fortunate that almost all the problems that arise in chemistry are capable of solution in terms of only four particles and one force. Unfortunately, although the phenomena are capable of explanation, there are often overwhelming practical and computational difficulties that prevent us obtaining quantitative results.

Protons and neutrons make up the nuclei of atoms. The mass of atoms lies largely in the nuclei which, though massive, are very small, around 10^{-15} m. The electrons which surround the nuclei possess a mass only 1/1836 that of a proton. The distance of the electrons from the nucleus, on the order of 10^{-10} m, determines the size of atoms and is typically expressed in picometres or as a fraction of a nanometre. However, the size of atoms is also often quoted in a more traditional unit, the Ångstrom unit (Å), which is equal to 10^{-10} m. For an atom that has no overall electrical charge, the number of electrons is equal to the number of protons in the nucleus. This number is known as the *atomic number*, Z, and determines the chemical properties of the atom. There are approximately 100 distinct types of atom and each defines a chemical element. Thus, for every possible atomic number from one to about 100, there exists an element whose atoms have in the nucleus that number of protons and which contain the same number of orbiting electrons.

1.3 The periodic table

The chemical elements show a diverse range of properties. About 80% are metals which have a shiny surface and are malleable, ductile and good conductors of electricity and heat. Apart from the broad categorisation of elements into metals and non-metals, elements can be grouped into families that exhibit very similar properties. One such family is noble (inert) gases, including helium (He) and argon (Ar). These are very unreactive. By contrast, the group of elements known as the alkali metals, which include sodium (Na) and potassium (K), are very reactive. They burn in air and react vigorously with water to release hydrogen.

Early in the 19th century chemists became aware that elements could be classified into these and other groups with very similar properties. However, it was not until 1869 that Dmitri Mendeleyev showed that the elements could be arranged in order of their atomic masses in such a way as to bring together all the elements with similar properties. This table was one of the greatest contributions to the understanding of chemistry in the late 19th century. The discovery of its theoretical basis was one of the most important contributions to our understanding of chemistry in the early 20th century. Mendeleyev was able to predict the properties of yet-undiscovered elements by noting gaps in his table. One of these he called ekasilicon, as it lay below silicon, and, some 15 years after his prediction, the element germanium (Ge) was discovered with properties uncannily close to those he predicted. The fact that the use of predicted properties enabled the discovery of a number of elements provided convincing evidence of the validity of his table. However, Mendeleyev's formulation was not entirely satisfactory. The use of atomic masses led to discrepancies which Mendeleyev boldly ignored and which were not resolved until it was realised that the important variable for classifying elements was not atomic mass, but *atomic number*, that is, the charge on the nucleus. The modern form of the table consists of rows of 2, 8, 8, 18, 18, 32 and 32 elements, each arranged so that the elements lie in columns, called *groups*, with others of similar physical and chemical properties (see p. 299). An understanding of the way that electrons behave in atoms enables us to predict these patterns of reactivity amongst the elements.

1.4 Isotopes

Atoms of the same element have the same number of protons in the nucleus but may have different numbers of neutrons. These different varieties are called *isotopes*. Chemical properties depend on the number of electrons surrounding the nucleus and much less on the overall mass of the atom. Therefore, isotopes have very similar chemical properties and are difficult to separate by chemical means. By analogy with the *atomic number*, Z, we may define a *mass number*, A, defined as the sum of the number of protons and neutrons in the nucleus, which is closely related to the mass of the atom. The nucleus of the normal isotope of hydrogen contains one proton, but other isotopes exist. The isotope with one neutron and one proton in its nucleus is called deuterium and the isotope with two neutrons and one proton is called tritium. The isotopes of other elements do not have individual names, and a systematic procedure for identification is used in which the symbol for the element carries the mass number as a superscript and the atomic number as a subscript. Therefore, the normal isotope of hydrogen is written $_1^1H$ and deuterium as $_1^2H$. Hydrogen consists very largely of its normal isotope, but some elements are mixtures with significant proportions of more than one isotope. For example, chlorine has the two isotopes $_{17}^{35}Cl$ and $_{17}^{37}Cl$ in the ratio of approximately three to one.

Chemists are largely concerned with the behaviour of stable isotopes, but some isotopes have a ratio of neutrons to protons in their nucleus which leads them to become unstable and to have only a short life. They usually disintegrate in one of two ways. Either they eject an alpha particle which is comprised of two neutrons and two protons (that is, the nucleus of the helium atom, $_2^4He$), or a beta particle, which is simply an electron formed by the conversion of a neutron into a proton and an electron. For example, normal hydrogen and deuterium are stable isotopes of hydrogen, but tritium decays emitting a beta particle to form a rare isotope of helium:

$$_1^3H \rightarrow {_2^3He^+} + e.$$

Similarly, the most abundant isotope of uranium decays by alpha-particle emission to form an isotope of the element thorium:

$$_{92}^{238}U \rightarrow {_{90}^{234}Th} + {_2^4He}.$$

When a radioactive nucleus disintegrates a great deal of energy can be released in the form of gamma radiation, a high-energy form of electromagnetic radiation, and in the kinetic energy of the particles produced. This energy comes from the conversion of tiny amounts of mass into energy according to the Einstein formula, $E = mc^2$. The conversion of 1 g of matter into energy, calculated by using this equation, is 9×10^{13} J; a factor 10^9 times greater than that released by burning the same mass of coal.

1.5 Molecules

Atoms rarely exist in isolation. In most cases, they link themselves to other atoms to form molecules. In 1803, John Dalton, observing that substances combined in simple ratios, proposed his quantitative atomic theory which became the basis for

Table 1.2. Valencies of elements in common molecules.

Element	Valencies	Compound	
hydrogen	1	H_2	hydrogen
		HCl	hydrogen chloride
		H_2O	water
oxygen	2	O_2	oxygen
		H_2O	water
		CO_2	carbon dioxide
nitrogen	3	NH_3	ammonia
	4	NO_2	nitrogen dioxide
	5	N_2O	nitrous oxide
chlorine	1	Cl_2	chlorine
		HCl	hydrogen chloride
		CCl_4	tetrachloromethane (carbon tetrachloride)
sodium	1	NaCl	sodium chloride
		NaOH	sodium hydroxide
sulfur	2	H_2S	hydrogen sulfide
	6	SF_6	sulfur hexafluoride
carbon	4	CH_4	methane
		C_2H_4	ethene

the systematic study of chemistry. The fact that molecules such as CH_4, C_2H_6 and C_2H_4 exist, but many other combinations of carbon and hydrogen atoms, for example, CH_7, do not, led to the understanding that each atom is capable of making only a limited number of chemical bonds with other atoms. These numbers are characteristic of each element and are termed the *valencies* of the element. They are largely determined by the group in the periodic table in which the element lies (Table 1.2).

Chemical bonding arises from the electrical charges on the atomic nuclei and on the electrons. If the electrons with their negative charges can occupy the space between two positively charged nuclei, the net effect is a strong electrical attraction between the nuclei and chemical bonding can occur, forming a molecule. The electrons act like glue and hold the nuclei together. In their absence, the two nuclei would repel each other. Not all atoms will bind together to form molecules. For example, if two argon atoms come together, only a weak attraction will occur which is insufficiently strong for a stable molecule, Ar_2, to be formed. Two types of situation can lead to bond formation. First, when the electrons can rearrange so that one atom has a net positive charge and the other a net negative charge. This is called *ionic bonding*. The other situation is when two atoms can share electrons and is called *covalent bonding*. To understand the tendency of atoms to form chemical bonds, we must find the rules to tell us when an atom might be expected to lose, gain or share electrons. These rules depend, to a large extent, on the number of outer electrons of the atom and the strength with which they are bound to the nucleus.

1.6   The mole

Matter is composed of atoms and molecules — the extremely small particles which are the basic building blocks of chemistry. 'Normal' quantities of substances that can be contained in a cup or weighed on a laboratory scale contain unimaginably large numbers of atoms, on the order of 10^{23}. Some impression of the number of molecules is given by the anecdote that, assuming a reasonable amount of mixing in the atmosphere, each time we breathe we inhale some molecules of Napoleon's (or any other historical person's) dying breath! Chemistry involves the study of atoms and molecules in these very large numbers and it is often inconvenient to deal with the properties of individual atoms and molecules, which are very small and unrelated to the magnitudes of properties we measure in the laboratory. To solve this problem, we use the concept of the *mole*. The mole is the fundamental unit of quantity of material. It provides a convenient way of scaling up molecular masses to those on the laboratory scale of measurement. A mole of substance is an amount that contains as many molecules of that substance as there are atoms in exactly 12 g of the ^{12}C isotope of carbon. This quantity, called the *Avogadro constant*, is $6.022 \times 10^{23} \, \text{mol}^{-1}$. We can have moles of not just molecules, but of ions, atoms, or any other particle. The mass of a mole of molecules of a substance is termed its *molar mass* and is usually expressed in units of g mol^{-1}. The *relative molar mass*, often loosely called the molecular weight of a substance, is the dimensionless ratio of the mass of a molecule to a unit of mass which is approximately the mass of a hydrogen atom. This unit, the *atomic mass unit* (amu), is more accurately defined as 1/12 the mass of an atom of ^{12}C and equals $1.6605 \times 10^{-27} \, \text{kg}$.

1.7   Waves

The roots of physical chemistry were established at the end of the 19th century when so-called *classical physics* was at its peak. They were based on the understanding of the properties of particles and waves which were then regarded as separate and distinct phenomena. Particles were seen as objects with mass, velocity and nonzero size which were, at any instant, located in a definite position. Waves, on the other hand, were seen as an undulation that moves forward carrying energy (though the substance oscillating does not itself move forward).

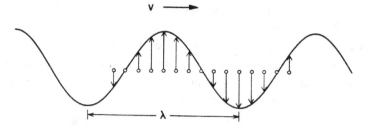

Fig. 1.2. Wave motion: the wave arises from the up-and-down motion of matter as shown. The distance the wave moves in unit time is its speed. The wavelength is the distance between successive peaks or troughs.

Waves are characterised by the wavelength, λ, the distance between repeats in the pattern, and the frequency, ν, the number of oscillations per second (Fig. 1.2). The reciprocal of the frequency is the period of the oscillations, τ,

$$\tau = 1/\nu.$$

The speed, v, of the wave motion is equal to the wavelength multiplied by the frequency of the oscillations, or, alternatively, divided by the period, and is given by

$$v = \lambda\nu = \lambda/\tau$$

It is a characteristic of wave motion that two or more waves can interfere producing either reinforced or reduced amplitudes at a given point. If two waves have the same amplitude at all points they are said to be *in phase* (Fig. 1.3). If they are in phase, two

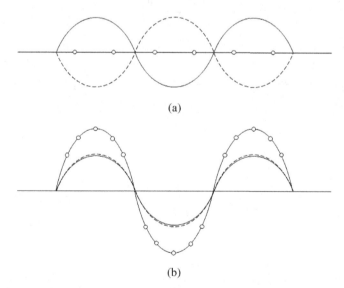

(a)

(b)

Fig. 1.3. The interference of two waves indicated by a solid line and a broken line. The resultant is identified by open circles: (a) the resultant is zero; (b) the signal is of enhanced intensity.

identical waves will add together to produce a wave of the same frequency but with twice the amplitude. On the other hand, if the waves are exactly half a wavelength out of phase, they *interfere* so as to cancel each other out and the net result is zero. At intermediate phase differences the amplitude lies between those two limits. When two waves of different frequency are superimposed more complex patterns are generated. Waves scattered from multiple sources can interfere to produce diffraction patterns. Such patterns, when produced by radiation of appropriate length, are used by chemists and physicists to determine the arrangement of atoms within molecules and the structure of solids and liquids.

1.8   Electromagnetic radiation

Electromagnetic waves play an important role in physical chemistry. In the 19th century, James Clerk Maxwell showed that electricity and magnetism were two aspects of the same electromagnetic force. Electromagnetic waves have an oscillating electric field at right angles to the direction of propagation and an oscillating magnetic field at right angles to both the direction of propagation and that of the electric field (Fig. 1.4). The properties of electromagnetic radiation depend very much on the wavelength (Fig. 1.5). This ranges from radio waves with a wavelength of many metres, through light waves ($\lambda \sim 10^{-7}$ m), to high energy X-rays and γ-rays, with wavelengths of less than 10^{-11} m. Electromagnetic waves move with the speed of light, c, and are often characterised by their wave number

$$\bar{v} = \frac{1}{\lambda} = \frac{v}{c}$$

usually expressed in units of cm^{-1}. Under certain conditions, electromagnetic radiation can be absorbed or emitted by atoms and molecules producing patterns called spectra. Microwave spectra, corresponding to frequencies of $\sim 10^{12}$ s^{-1},

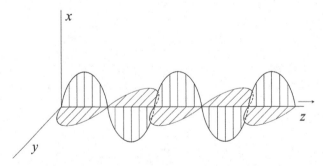

Fig. 1.4. Electromagnetic radiation consists of an oscillating electric field and an oscillating magnetic field. The magnetic field component is at right angles to the electric component. Normally, the oscillations can occur in all directions in the x–y plane and the radiation is then said to be unpolarised.

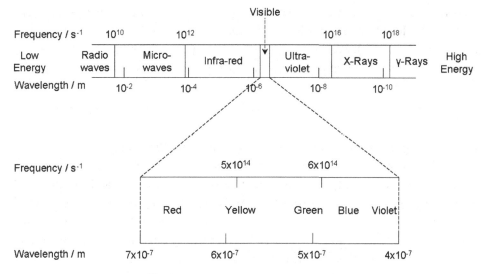

Fig. 1.5. The electromagnetic spectrum.

provide information about molecular rotations. Molecular vibrations, which take place much faster than rotations, typically in $\sim 10^{14}$ s^{-1}, can be studied in the infrared region. Electronic motion in atoms gives rise to spectra in the visible and ultraviolet regions. To interpret the spectra we shall need an understanding that cannot be provided by classical physics.

1.9 The perfect gas

The perfect gas plays an important role in physical chemistry. It provides a model for the behaviour of real gases which is followed reasonably accurately at low and moderate pressures. It also provides a state that can be used as a reference to which the behaviour of real substances can be related. A perfect gas may be defined as one that satisfies the conditions that pressure, volume and temperature are related by the equation

$$PV = nRT,$$

where P is the pressure, V volume, n the number of moles of gas and R is the gas constant. T is the temperature expressed on the absolute temperature scale. The energy of the perfect gas depends only on its temperature and not on the pressure or volume, behaviour that can be shown to be a consequence of the perfect gas equation (Section 2.2). In a perfect gas, the molecules are assumed to take up no space and so, in a mixture of such gases, each of the gases present acts as if it were alone in the container. The total pressure is the sum of the pressures each gas would exert if it were alone in the same volume. For a mixture of n_A moles of perfect gas A and n_B moles of perfect gas B, the *partial pressure* of substance A is defined as $P_A = P n_A / (n_A + n_B)$. We can write this as $P_A = x_A P$, where $x_A = n_A / (n_A + n_B)$ is the *mole fraction* of component A. The total pressure P is the sum of all partial pressures

$$P = P_A + P_B.$$

The perfect gas equation provides us with, amongst other things, a definition of temperature. For a fixed quantity of a perfect gas (and real gases approximate this limit quite accurately under normal conditions and follow it exactly in the limit where the pressure tends to zero), we can write $P_1 / P_2 = T_1 / T_2$. As P is zero when T is zero, we need only one point to define a scale of temperature. We choose the so-called *triple point* of water at 273.16 K, a point very close to the normal melting point of ice, which is at 273.15 K (Section 9.1).

1.10   Chemical equilibrium

Many chemical reactions are able to proceed in a forward or in a reverse direction depending on the conditions. They do not proceed to completion but reach a position of equilibrium after which no further change occurs. A well-studied example is the reaction of hydrogen and iodine to produce hydrogen iodide, which we can write as

$$H_2 + I_2 \rightleftharpoons 2HI.$$

The two oppositely-directed arrows indicate the reversibility of the reaction. In 1867, Guldberg and Waage proposed the *law of mass action* which recognised that the position of equilibrium in a reaction could be defined in terms of the concentrations of the reactants and products. For a general reaction between a molecules of substance A and b molecules of substance B to produce l molecules of substance L and m molecules of substance M,

$$aA + bB \rightleftharpoons lL + mM,$$

we can define a function, Q, the *reaction quotient*, by

$$Q = \frac{[L]^l [M]^m}{[A]^a [B]^b},$$

where the square brackets indicate the concentrations and $a, b, l,$ and m are the *stoichiometric coefficients*. As the reaction proceeds, Q changes until, at equilibrium, it reaches a value that is a constant at any given temperature called the *equilibrium constant*, K_c, defined by

$$K_c = \left(\frac{[L]^l [M]^m}{[A]^a [B]^b} \right)_{eq},$$

where all the concentrations are at their equilibrium values. The concentrations in equations such as this are usually expressed as dimensionless ratios, c/c^0, where c^0 is the unit of concentration employed. Equilibrium constants defined in this way are dimensionless. A unit of concentration frequently used is that of unit *molarity*, 1 mol dm^{-3}. The concentrations of solutions are also expressed in terms of *molality*. The unit of molality is 1 mole of solute per kilogram of solvent, mol kg^{-1}. As 1 dm^3 of water under normal conditions has a mass of approximately 1 kg (0.997 kg at 298 K), the molarity and molality of aqueous solutions can, for most practical purposes, be taken as equal.

If Q is greater than K_c, the reaction will approach the equilibrium position from the right-hand side of the equation. If Q is less than K_c, the reaction will proceed to the equilibrium position from the opposite direction. In each case, the reaction will attain the same position of equilibrium defined by K_c. (A more general treatment of equilibrium constants will be introduced in Chapter 8.)

For reactions in the gas phase, we can define the position of equilibrium in terms of the partial pressures of the reactants and products. (Partial pressures are defined in Section 1.9.)

$$K_P = \left(\frac{P_L^l P_M^m}{P_A^a P_B^b} \right)_{eq}$$

In this equation, the partial pressures are expressed in terms of the dimensionless ratio, $P/(1 \text{ bar})$, so that K_P is, like K_c, dimensionless.

When solids or pure liquids take part in an equilibrium reaction, their concentration is assumed to be constant. For example, the concentration of a solid in solution does not depend on the quantity of material present, only on the solubility of the solid substance. The same considerations apply in the vapour phase, where the concentration of a liquid or solid substance depends on the vapour pressure and not the quantity present. For the equilibrium

$$3Fe(s) + 4H_2O(g) \rightleftharpoons Fe_3O_4(s) + 4H_2(g),$$

we find that the equilibrium constant, K_P, is given by $K_P = (P_{H_2}/P_{H_2O})^4$ and that the ratio (P_{H_2}/P_{H_2O}) is a constant at any given temperature. The quantities of the two solid materials present do not affect the equilibrium position.

The manner in which the position of equilibrium responds to changes in the external conditions, such as temperature and pressure, can be deduced, qualitatively, from *Le Chatelier's principle*. This can be expressed as follows: *if any change is imposed on a system at equilibrium, the position of equilibrium will shift to counteract, at least in part, that change*. Thus, if the forward reaction in a chemical equilibrium gives out heat (is *exothermic*), then lowering the temperature will promote the reaction, leading to an increase in the quantity of products present in the equilibrium mixture. Conversely, if the forward reaction absorbs heat (is *endothermic*), reducing the temperature will lead to a decrease in product quantity. In a similar manner, if a reaction proceeds with an increase in volume, increasing the pressure will reduce the extent of reaction. Chemical thermodynamics, introduced in Chapters 7 and 8, enables us to predict the response of equilibrium constants to changing temperature and pressure in a quantitative way. Le Chatelier's principle has been widely applied in disciplines other than chemistry, including biology and economics.

1.11   Ionic equilibria

Many chemical compounds separate into ions when dissolved in a solvent such as water, in which the electrostatic forces between the oppositely charged ions are greatly reduced. Such electrolyte solutions are characterised by their ability to conduct electricity. The degree of dissociation of a substance in solution determines its strength as an electrolyte. Many substances, such as NaCl, HCl and NaOH, are almost completely dissociated into ions in dilute aqueous solution. Other compounds, such as ethanoic (acetic) acid, are only very sparingly dissociated, with a degree of dissociation that is strongly dependent on their concentration in solution. For the general case, in which a compound dissociates into two ions, we can write

$$AB \rightleftharpoons A^+ + B^-$$
$$(1 - \alpha) \qquad \alpha \qquad \alpha,$$

where the fraction of AB dissociated into ions is α, the *degree of dissociation*. If the total concentration of the compound is c, the concentration of each of the ions is αc, and that of the unionised compound $(1 - \alpha)c$. We can define a dissociation constant K_d by

$$K_d = \frac{[A^+][B^-]}{[AB]} = \frac{(\alpha c)(\alpha c)}{(1 - \alpha)c} = \frac{\alpha^2 c}{(1 - \alpha)}$$

This relationship is known as *Ostwald's dilution law*. For a weak electrolyte, $\alpha \ll 1$ and we can use the approximation $K_d \approx \alpha^2 c$, giving $\alpha \approx (K_d/c)^{1/2}$. We find for ethanoic acid at a 0.001 mol dm^{-3} concentration in solution at 25°C, $\alpha = 0.13$, giving $K_d = 1.9 \times 10^{-5}$. The dilution law predicts a degree of dissociation, $\alpha = 0.004$ for a molar (1 mol dm^{-3}) solution of ethanoic acid, in reasonable agreement with the experimentally-determined value. When dealing with the dissociation of acids in solution, we normally express the dissociation constant as the *acidity constant* or the acid dissociation constant, K_a.

In a saturated solution of a solid electrolyte, the solid is in equilibrium with the ions in solution. For example, for a solution of solid silver chloride,

$$Ag\,Cl(s) \rightleftharpoons Ag^+ + Cl^-.$$

The product $[Ag^+]\,[Cl^-]$ is called the *solubility product* of silver chloride.

Acids and bases constitute very important classes of electrolyte. Acids are generally defined as substances that give rise to a hydrogen ion when dissociated in solution,

$$HA \rightleftharpoons H^+ + A^-.$$

Strictly, since hydrogen ions are unlikely to exist in solution other than in a hydrated form, they are better represented by H_3O^+, but we will continue to write H^+ in the interests of simplicity. The strength of an acid is reflected in its acidity constant, K_a.

Bases were traditionally defined as substances that dissociate to liberate hydroxyl ions in solution,

$$BOH \rightleftharpoons B^+ + OH^-.$$

The strength of such a base is defined in terms of the dissociation constant K_b, defined as

$$K_b = \frac{[B^+][OH^-]}{[BOH]}.$$

Water dissociates very slightly to produce $[H^+]$ and $[OH^-]$ ions. At room temperature, the degree of dissociation is 10^{-7} and we can define the dissociation constant of water, termed the *ionic product* of water, K_w, by

$$K_w = [H^+][OH^-] = 10^{-14}.$$

Hydrogen ion concentrations are most conveniently defined in terms of pH, where $pH = -\log[H^+]$. In pure water, the value of the pH is given by

$$pH = -\log 10^{-7} = +7.$$

Acid solutions have pH values less than seven. For example, a 0.1 mol dm^{-3} solution of hydrogen ions would have a pH of 1. Conversely, bases have pH values greater than seven. A $0.1 \, mol \, dm^{-3}$ aqueous solution of a strong base for which $[OH^-] = 0.1$ has a hydrogen ion concentration given by

$$[H^+] = K_w/[OH^-] = 10^{-14}/0.1 = 10^{-13} \text{ and consequently a pH of 13.}$$

Brønsted and Lowry independently proposed a more general definition of acids and bases. They suggested that any compound that can transfer a proton to another compound be defined as an acid and that compounds that can receive protons be defined as bases. All acid-base reactions involve an acid and its conjugate base which can receive the proton. For example, the ammonium ion can be regarded as an acid in the reaction

$$NH_4^+ \rightleftharpoons NH_3 + H^+,$$

where NH_3 is the conjugate base. The Brønsted–Lowry definition permits the concept of acid-base reactions to be extended to non-aqueous solutions.

A further extension of the concept of acids and bases was due to G. N. Lewis, who defined acids as substances that can receive a lone pair of electrons from another molecule. This definition includes, as well as more conventional acids, substances such as BF_3. Substances defined under this convention as acids, but which do not release protons, are termed Lewis acids.

1.12 The next steps

In this chapter, we have summarised the basic concepts underpinning physical chemistry.

In particular, we have identified two important principles. First, that *energy is not continuous and may only take on certain fixed values*, that is, it is *quantised*. The consequences of this seemingly simple fact are widespread and dramatic. In Chapter 2, we will outline the behaviour of energy in classical physics. In Chapter 3, we elucidate the consequences of the quantisation of energy and of the wave-like behaviour of particles. This is the subject of *quantum mechanics*.

The second principle is the tendency of systems to move to a condition of the *maximum randomness* (other things being equal). This is the subject of *chemical thermodynamics* which is investigated in Chapter 7 and which enables us to understand chemical equilibrium, the subject of Chapter 8. Chapters 9 and 10 describe the behaviour of matter in its various states (gas, liquid and solid) and in mixtures. Finally, in Chapter 11, we examine the rates of chemical reactions and investigate how they can be understood in terms of the molecular properties of the reactants and products of a reaction.

2
Energy

Understanding the energy of chemical systems is of crucial importance to chemists. The properties of atoms and molecules are determined, to a large extent, by the magnitudes of the various forms of energy they contain. The energy changes accompanying chemical processes are a major (but not the only) factor in determining the direction in which reactions can proceed. An understanding of energy was not easily obtained and, although the first tentative steps were made in the 17th century, it was not until the 19th century that the concept was fully established. Nowadays, the most frequently-employed general definitions of energy relate it, somewhat unhelpfully, to the capacity to raise weights. While being aware of such definitions, we will introduce energy in terms of more relevant definitions.

2.1 Kinetic and potential energy

The energy possessed by bodies is of two types. The energy which arises by virtue of the motion of a body is referred to as the *kinetic energy* and is defined by the equation

$$E_K = \frac{m\mathrm{v}^2}{2} = \frac{p^2}{2m},$$

where m is the mass, v the velocity and p the linear momentum, $m\mathrm{v}$, of the body in motion.

In addition to their kinetic energy, bodies can possess *potential energy* due to the forces that act on them by virtue of their position. The most common example of this is the energy that arises from gravitational forces. This energy, which depends on the height of the body in the Earth's gravitational field, is termed the gravitational energy, V, and is given by $V = mgh$, where g is the acceleration due to gravity, which depends somewhat on location but is approximately 9.81 m s^{-2}. The kinetic energy of a body at rest is zero. However, the zero of potential energy is arbitrary and can be set at a convenient point. Thus, it is a common convention to regard the gravitational potential energy at the surface of the Earth as zero.

More generally, we can write that the change in the potential energy of a body experiencing a force $f(x)$, which depends on its position x, is

$$V(x_2) - V(x_1) = - \int_{x_1}^{x_2} f(x)dx.$$

We can also define force by

$$f(x) = -\frac{dV(x)}{dx}.$$

The potential energy arising from gravitational forces plays an almost negligible role in determining the structures and processes of interest to chemists. By far the most important forces in chemical systems are the electromagnetic forces, called the Coulombic forces, which arise between electrically charged bodies.

The total energy of molecules arises from the potential energy and kinetic energy contributions. The kinetic energy that results from the motion of the centre of mass of an atom or molecule is called *translational energy*. Another contribution to molecular energy arises from the *rotation of molecules*. The energy of a rigid diatomic molecule with two atoms of mass m a distance r apart, rotating with an angular velocity ω, is given in classical mechanics by $E_{rot} = I\omega^2/2$, where I is the moment of inertia. This is defined as $I = \sum m_i r_i^2$, where m_i is the mass of atom i, r_i the distance of atom i from the centre of mass of the molecule and the sum is over all the atoms in the molecule. A *vibrating molecule* undergoing simple harmonic motion has a force on the atoms directly proportional to the distance from the equilibrium position, the amplitude of the vibration, A, (Hooke's law). In classical mechanics, the vibrating molecule has a total energy proportional to the square of the maximum amplitude, A_{max}, $E_{vib} = KA_{max}^2$, where K is the force constant. The vibrational energy arises from both potential and kinetic contributions. At the turning point, i.e. at the maximum amplitude, the kinetic energy is zero and the potential energy will be a maximum. At the centre point of the vibration, the potential energy is equal to zero, as there is no net force on the vibrating body, and both the velocity and the kinetic energy are at a maximum. The total energy of the oscillator is constant.

The most substantial contribution to the energy of atoms and molecules arises from the interactions of the electrons with the nucleus. This is termed *electronic energy*. Therefore, the total energy of an isolated molecule can be written

$$E = E_{trans} + E_{rot} + E_{vib} + E_{elec}.$$

We will investigate each of these contributions in some detail and discover that classical physics provides only a very limited understanding of how these energies arise.

2.2 Kinetic theory of gases

The behaviour of a perfect gas, a gas that follows the equation $PV = nRT$, can be understood if we regard the gas molecules as points of mass m, confined to a box of side a (Fig. 2.1). We assume that the molecules exert no forces on one another, are in constant motion, and that the only energy in the system is kinetic. We also assume that when they collide with the walls of the container, the collisions are perfectly elastic and involve no loss of energy.

Consider one molecule and let the molecule have a velocity v with components v_x, v_y and v_z. These components are the velocity of the molecule in the x, y and z directions, respectively. For the x direction, the molecule will hit the walls perpendicular to this direction at time intervals defined by the distance between the walls divided by the velocity, a/v_x. On each collision, the momentum of the molecule will change from $+mv_x$ to $-mv_x$, a total change of $2mv_x$. Therefore, the total change of momentum at the wall per unit time is $2mv_x(v_x/a) = 2mv_x^2/a$. The rate of change of momentum is, by Newton's second law, the force that the molecule exerts on the wall. This force acts on the two walls perpendicular to the x direction, which have a total area of $2a^2$. Since pressure is defined as the force per unit area, we have, for the pressure exerted by a single molecule,

$$P_x = \left(\frac{2mv_x^2}{a}\right)\left(\frac{1}{2a^2}\right) \quad \text{and} \quad P_x = \frac{mv_x^2}{V},$$

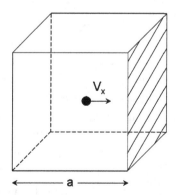

Fig. 2.1. A gas molecule with velocity v_x in a cubic box of side a.

as a^3 is equal to the volume V. If the box contains N molecules with the *average of the square of the velocity component* in the x direction, $\overline{v_x^2}$, we obtain

$$P_x = \frac{Nm\overline{v_x^2}}{V}$$

Gases are isotropic, that is, their properties are the same in all directions, and the pressures on all walls are identical. Thus,

$$P = \frac{Nm\overline{v_x^2}}{V} = \frac{Nm\overline{v_y^2}}{V} = \frac{Nm\overline{v_z^2}}{V}$$

since $\overline{v_x^2} = \overline{v_y^2} = \overline{v_z^2}$

The total speed of a molecule, v (which can be in any direction) is related to the velocity components by

$$v^2 = v_x^2 + v_y^2 + v_z^2 = 3v_x^2$$

and $\overline{v_x^2} = \overline{v^2}/3$. Therefore, $P = Nm\overline{v^2}/3V$, where $\overline{v^2}$ is the average value of the square of the speeds and its square root is termed the *root-mean-square speed*, c. As the total kinetic energy of the molecules is $E_K = Nmc^2/2$, we obtain

$$PV = \frac{2}{3}\left(\frac{Nmc^2}{2}\right) = \frac{2E_K}{3}$$

The perfect gas equation tells us that $PV = nRT$. Therefore, we obtain, for the kinetic energy of the gas molecules, $E_K = 3RT/2$. Temperature can be directly related to the kinetic energy of the molecules. Using the relation

$$PV = \frac{Nmc^2}{3} = nRT,$$

we can calculate the root-mean-square speed of gas molecules, c, from $c = (3RT/M)^{1/2}$ since, for one mole, Nm is equal to the molar mass, M. The speed of molecules depends only on the temperature and on the mass. It increases as the square root of the temperature and is inversely proportional to the square root of the molar mass. For nitrogen at room temperature (298 K) (and with the molar mass expressed in kilograms to be consistent with the SI units), we obtain $c = (3 \times 8.3 \times 298/0.028)^{1/2} = 515\,\text{m s}^{-1}$. As the kinetic energy of all species of molecules is the same at any given temperature, lighter molecules move faster than heavier ones. The root-mean-square-speed of hydrogen molecules at 298 K is $1920\,\text{m s}^{-1}$, whereas the value for sulfur hexafluoride (molar mass 146) is $226\,\text{m s}^{-1}$ at the same temperature. We see that molecules move at a high velocity, of the order of $10^2 - 10^3\,\mathrm{m\,s^{-1}}$ — comparable with, for example, the speed of a bullet from a rifle, a jet aeroplane or, more relevantly, the speed of sound in air under normal conditions, $340\,\mathrm{m\,s^{-1}}$.

Not all molecules of the same species travel with the same velocity. The distribution of molecular velocities in a gas is determined by two factors. The first is that fewer molecules have high velocities since the kinetic energy falls off according to the factor $\exp(-m\mathrm{v}^2/2kT)$. (We will investigate this important expression, known as the Boltzmann factor, in Chapter 7.) The second is due to the fact that the speed in three dimensions arises from the velocity components in the x, y and z directions $\mathrm{v}^2 = \mathrm{v}_x^2 + \mathrm{v}_y^2 + \mathrm{v}_z^2$. Because of the number of ways v can be made up, the probability of any value of v lying between v and $\mathrm{v} + d\mathrm{v}$ is proportional to $4\pi\mathrm{v}^2\,d\mathrm{v}$. (This is the volume of a spherical shell of thickness dv and of radius v.) This distribution favours higher values of v. The resulting molecular speed distribution is given by

$$\frac{dn(\mathrm{v})}{n} = 4\pi\left(\frac{m}{2\pi kT}\right)^{3/2}\exp\left(\frac{-m\mathrm{v}^2}{2kT}\right)\mathrm{v}^2 d\mathrm{v},$$

where $dn(\mathrm{v})/n$ is the fraction of molecules whose speed lies in the range v to $\mathrm{v} + d\mathrm{v}$. This expression was derived by James Clerk Maxwell in 1860 and is generally known as the *Maxwell distribution of molecular speeds*. It can be integrated to obtain the result that the *average speed* $\bar{\mathrm{v}} = (8kT/\pi m)^{1/2} = (8RT/\pi M)^{1/2}$. We note that the average speed of molecules is not the same as the root-mean-square speed, c, and we find $\bar{\mathrm{v}} = (8/3\pi)^{1/2}c = 0.92c$. The average speed of molecules, like that of the root-mean-square speed, depends only on the temperature and their mass. At low temperatures, the distribution is narrow with a maximum at a low speed. At higher temperatures, the distribution becomes broader and the maximum is located at higher values of the speed (Fig. 2.2).

When a gas under pressure passes through a small orifice, a process known as *effusion*, its rate of passage is determined by the average speed of the molecules. This phenomenon can be used to estimate the molecular masses of gas molecules.

2.3   Equipartition of energy

We have seen that the kinetic energy of a gas molecule can be regarded as made up of three independent contributions

$$\frac{m\overline{\mathrm{v}_x^2}}{2} + \frac{m\overline{\mathrm{v}_y^2}}{2} + \frac{m\overline{\mathrm{v}_z^2}}{2}.$$

The total kinetic energy arising from these contributions is $3RT/2$, where R is the gas constant and T the temperature. As the gas is isotropic, each component, in the x and y and z directions, will contribute $RT/2$ to the total energy. Each of these

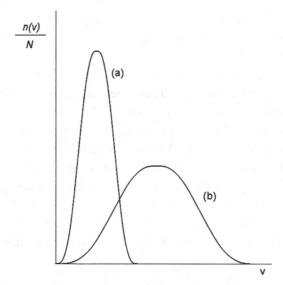

Fig. 2.2. The fraction of molecules with speed, v as a function of speed (arbitrary units): (a) low temperature; (b) high temperature.

contributions is termed a *degree of freedom*. In classical physics, each degree of freedom contributes a similar amount to the energy. Thus, each rotational mode will also contribute $RT/2$ to the energy of the molecule. A linear molecule has only two moments of inertia and, consequently, only two rotational degrees of freedom. Therefore, rotation would contribute RT to the energy. A non-linear molecule has three moments of inertia and three degrees of freedom and would be expected to contribute $3RT/2$. Each vibrational degree of freedom contributes *not* $RT/2$, but RT, as vibrational motion contributes two degrees of freedom, one by virtue of the kinetic energy of the vibrational motion and the other due to the potential energy arising from the restoring force.

If we add the various contributions to the molar energy of a diatomic molecule (other than its electronic energy), we obtain

$$E = E_{\text{trans}} + E_{\text{vib}} + E_{\text{rot}}$$
$$= \frac{3RT}{2} + RT + RT = \frac{7RT}{2}.$$

Non-linear polyatomic molecules have three rotational degrees of freedom and $(3n - 6)$ vibrational modes, where n is the number of atoms in the molecule. We can understand how this number arises if we recognise that every atom can move in three dimensions and, consequently, has three degrees of freedom. If three degrees are assigned to translational motion and three to rotation, then the remaining $(3n - 6)$

degrees of freedom must be vibrations. A non-linear diatomic molecule such as water will have three degrees of translational freedom, three degrees of rotational freedom and three degrees of vibrational freedom, and the principle of equipartition of energy would predict a total energy contribution from these sources of $6RT$.

For an atomic solid, each atom can be regarded as having three degrees of vibrational freedom and we would predict that the total molar energy would be $3RT$. The assumption of the equipartition of energy allows us to predict the heat capacities of molecules and of atomic solids. When a small quantity of energy is added to a substance, its temperature rises and the *heat capacity* is defined as the ratio of the quantity of energy added to the temperature increase, $C = (dE/dT)$. This would suggest that the heat capacity of atomic solids is $3R$ or approximately $25\,\mathrm{J\,K^{-1}mol^{-1}}$. For many solids at normal temperatures this was found to be a good approximation and is known as Dulong and Petit's law.

For a diatomic molecule with three degrees of translational freedom, two of rotation and one vibrational mode, a heat capacity of $7R/2$, approximately $29\,\mathrm{J\,K^{-1}mol^{-1}}$, is indicated. (The reasons why we can ignore the contribution of the electronic energy to the heat capacity will be made clearer in following chapters.) The assumption of the equipartition of energy is often a good approximation at high temperatures, but under many circumstances, particularly at low temperatures, it fails drastically. This represents a serious failure of simple mechanics and is one of the observations that confirmed that classical physics could not provide a satisfactory explanation of many molecular phenomena.

2.4 Heat and work

Before proceeding we must remind ourselves that energy can be transferred from one system to another in one of two different ways. *Heat* is the transfer of energy that results from temperature differences. When two bodies at different temperatures are brought into contact, heat will flow until the temperature gradient disappears and the two bodies reach the same temperature and thermal equilibrium is achieved. *Work*: A formal definition of work states that it is *always completely convertible into the lifting of a weight*. We will find it easier to understand if we recognise that, as chemists, we will usually only meet two types of work. By far the most common is that work which a system does by expanding, pushing back atmospheric pressure, or, conversely, the energy it gains when its volume is reduced under atmospheric pressure. We give work done *by* the system a negative sign since, by doing that work, the system loses energy.

$$w = -\int F \mathrm{d}L.$$

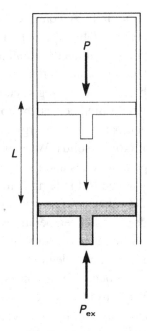

Fig. 2.3. Expansion of a gas against an external pressure.

The force on the piston in Fig. 2.3 is the pressure times the area, and, thus,

$$w = -\int (P_{ex}A)dL = -\int P_{ex}dV.$$

If the external pressure, P_{ex}, is always adjusted to be equal to the internal pressure of the gas within the cylinder during the expansion, the system does the maximum possible amount of work. (The significance of studying systems that do the maximum amount of work will be explained in Chapter 8.) The other commonly-occurring example of work that is important in physical chemistry is electrical work. This arises when a charge Q is moved through an electric potential difference Φ.

$$w = -\int \Phi dQ$$

This is the work that can be obtained from electrochemical cells. Unless some electrochemical process is indicated, we will assume that the only work done by the physicochemical systems we are studying is the so-called *PV*- work; the work of expansion against the prevailing atmospheric pressure.

We need to note that when transferred, both heat and work change the energy of the system. Heat and work are simply ways in which energy is transferred. The total energy change of a system, ΔU, is the net result of heat absorbed, q, and work done *on* the system, w (or, of course, if energy is lost, the heat evolved and the work done *by* the system),

$$\Delta U = q + w.$$

In going from one state to another, the energy change will always be the same but it may result from different quantities of heat and work being transferred to the system. Work and heat are fundamentally different in that work can always be used to generate an equivalent amount of heat, but the converse is not true. In general, heat can only be used to produce part of its energy in the form of work. In fact, this is a most important clue to the fact that we need an additional factor, other than energy, to define the stability of chemical systems. This will be the subject of Chapter 7.

2.5 Conservation of energy — The First Law

The fact that energy is conserved in physical processes and cannot be created or destroyed is commonly known as the *First Law of Thermodynamics*. This states that *the algebraic sum of all energy changes in an isolated system is zero*. An isolated system is one that cannot exchange energy or matter with its surroundings. Any transfer of energy from a system to its surroundings must be accompanied by a change in the energy contained in the system, its *internal energy, U*.

$$\Delta U = U_B - U_A = q + w$$

We could go from state A to state B in different ways in which different values of q and w would occur as shown in Fig. 2.4. However, the net result $(U_A - U_B)$ must always be the same. If that were not the case, we could go round the loop taking Path I (A $\rightarrow$ B) followed by Path II (B $\rightarrow$ A) in the opposite direction such that $\Delta U_I \neq \Delta U_{II}$ and the system would gain or lose energy, thus contradicting the First Law of Thermodynamics.

2.6 State functions

Internal energy, U, is one of a number of so-called *state functions* which depend only on the state of the system. Others include pressure, temperature and volume. Some, like P and T, do not depend on how much material is in the system. These are called *intensive* properties. Others, such as U and V, depend directly on the quantity of material under consideration. These are called *extensive* properties.

We can illustrate the properties of state functions by considering two cities A and B (Fig. 2.5). The distance we measure on a journey from A to B will depend

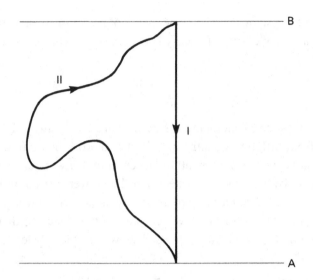

Fig. 2.4. Two possible paths (I and II) between states A and B.

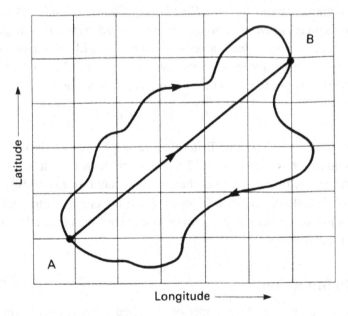

Fig. 2.5. Latitude and longitude as 'state functions'.

on the route we take but, however we go, the change in latitude and the change in longitude will be the same. Latitude and longitude are similar to the thermodynamic state functions and depend only on the positions of the cities A and B. The distance travelled depends on the choice of route and is not analogous to a state function.

In this respect it is similar to heat or work, which depend on the path we take between any two states. For any journey that starts and finishes at the same point, the changes in the latitude and longitude must add up to zero. In the same way, the changes in any state function over any complete cycle must also be zero.

These properties of state functions, which are very important in chemistry, can be expressed mathematically:

(1) If we integrate a state function between two limits the integral, for example $\Delta U = \int_A^B dU$, must have a definite value. This is just a mathematical statement of the property we have identified above. dU is said to be an exact differential. The integral of dU over any closed path is equal to zero.

(ii) We can write exact differentials

$$dU = (\partial U/\partial x)_y dx + (\partial U/\partial y)_x dy,$$

where x and y are variables which determine the value of U. $(\partial U/\partial x)_y$ is the rate of change of U with x when y is kept constant. It is known as a *partial differential coefficient*.

(iii) When a state function is differentiated with respect to two variables, the order of differentiation is immaterial. Thus,

$$\left[\frac{\partial (\partial U/\partial x)_y}{\partial y}\right]_x = \left[\frac{\partial (\partial U/\partial y)_x}{\partial x}\right]_y.$$

When a function or its differential coefficients satisfy any of these criteria, the function is a state function. The relationships between those state functions which are the properties of chemical systems are the principal concern of chemical thermodynamics. These relationships enable important quantities to be determined from others that are more easily measured in the laboratory.

2.7 Enthalpy

When we conduct experiments in the laboratory, the apparatus is usually not harnessed to do work. In fact, the only work that is done is PV- work, i.e. expansion (or contraction) against atmospheric pressure. If heat is added to a system that does no work, other than that against the atmosphere, taking it from state A to state B we have

$$\Delta U = U_B - U_A = q + w = q_P - P\Delta V = q_P - P(V_B - V_A),$$

where q_P is the heat added to the system at constant pressure. Thus, we can write

$$q_P = (U_B + PV_B) - (U_A + PV_A).$$

We define *enthalpy* as

$$H = U + PV.$$

Thus, $\Delta H = q_P$ and $\mathrm{d}H = (\mathrm{d}q)_P$.

Enthalpy is also a state function and ΔH has a definite value for any particular change under all conditions. However, it is only equal to the heat absorbed by the system, q, if the process is carried out at constant pressure. ΔH and ΔU are usually quite similar for changes involving solids and liquids, as the associated volume changes are usually small. However, for gases, there can be a distinct difference. For a gaseous reaction in which there is a change of Δn moles of perfect gases accompanying the reaction,

$$\Delta_r H = \Delta_r U + \Delta n RT.$$

At 298 K, $RT = 2.5\,\mathrm{kJ\,mol^{-1}}$ — not a negligible difference in terms of the energy changes typically observed for chemical processes.

We define the heat capacity of a substance by the temperature rise that is observed when a quantity of energy is added to the system, $C = (\mathrm{d}q/\mathrm{d}T)$. If heat is added at constant volume and the system does no work, so that $w = 0$, then $\mathrm{d}q = \mathrm{d}U$ and the heat capacity is given by

$$C_V = \left(\frac{\partial q}{\partial T}\right)_V = \left(\frac{\partial U}{\partial T}\right)_V.$$

If, on the other hand, the heat is added at constant pressure, then the heat capacity at constant pressure is given by

$$C_P = \left(\frac{\partial q}{\partial T}\right)_P = \left(\frac{\partial H}{\partial T}\right)_P.$$

As ΔH and ΔU are very similar for solids, C_P and C_V are also similar, but, for n moles of perfect gas,

$$H = U + nRT \quad \text{and} \quad C_P = C_V + nR.$$

The variation of the enthalpy change accompanying a reaction or a physical change with temperature can be calculated from a knowledge of the heat capacities of the reactants and products. Thus,

$$\left(\frac{\partial H}{\partial T}\right)_P = C_P \quad \text{and} \quad \left(\frac{\partial \Delta H}{\partial T}\right) = \Delta C_P.$$

When integrated, we obtain

$$\Delta_r H_2 - \Delta_r H_1 = \int_{T_1}^{T_2} \Delta_r C_P \mathrm{d}T.$$

If we make the approximation that the heat capacities are independent of temperature,

$$\Delta_r H_2 - \Delta_r H_1 = \Delta_r C_P (T_2 - T_1).$$

These relationships are called *Kirchhoff's equations* and enable us to determine the variation of the enthalpy change with temperature.

2.8 Hess's law

The fact that energy is conserved can be very helpful when trying to determine the energy change accompanying chemical reactions. This application of the conservation of energy was first recognised by Hess in 1840. Its value is that it enables us to calculate the energy change accompanying reactions for which a direct measurement would be difficult to carry out, from a knowledge of the energy changes accompanying other reactions which are more experimentally accessible. Since we are most often concerned with reactions that occur at constant pressure, we employ the state function enthalpy rather than the internal energy.

If we wish to know the enthalpy change accompanying a reaction, we can often conveniently determine this quantity, using *Hess's law*, from the enthalpies of formation of the reactants and the products. Let us take, as an example, the general reaction

$$AB + CD \rightarrow AC + BD$$

for which we write the enthalpy of reaction as $\Delta_r H$. We can obtain $\Delta_r H$ from the following enthalpies of formation:

$$
\begin{aligned}
AB &\rightarrow A + B & -\Delta_f H(AB) \\
CD &\rightarrow C + D & -\Delta_f H(CD) \\
A + C &\rightarrow AC & \Delta_f H(AC) \\
B + D &\rightarrow BD & \Delta_f H(BD)
\end{aligned}
$$

Summing these reactions gives

$$AB + CD \rightarrow AC + BD$$

and Hess's law tells us that the enthalpy change accompanying this reaction will be given by the sum of the enthalpy changes of the component reactions

$$\Delta_r H = \Delta_f H(AC) + \Delta_f H(BD) - \Delta_f H(AB) - \Delta_f H(CD)$$

$$\Delta_r H = \Delta_f H(\text{products}) - \Delta_f H(\text{reactants}).$$

This result reflects the fact that, when a chemical reaction takes place, the elemental composition is unchanged — the elements are merely redistributed between the compounds. This provides an important route to the calculation of the enthalpies of reaction. The *enthalpies of formation* from their elements of most common chemical compounds in their standard states, $\Delta_f H^0$, have been determined and are available in tables (Appendix 1). The standard states of compounds and elements are defined as the pure substance at one bar pressure and at the temperature specified — most commonly 298.15 K. The enthalpies of formation of the elements, *in their most stable form* and in the standard state, are, by definition, zero.

We can now apply the relations we have developed to investigate the *water–gas reaction* in which steam is passed over heated charcoal or coke,

$$C(s) + H_2O(g) \rightarrow CO(g) + H_2(g).$$

This is an important process for synthesising a gas that is used as an industrial fuel. Applying Hess's law to determine the enthalpy change accompanying the reaction we obtain, with all reactants and products in their standard states at 298 K (remembering that the standard enthalpy of formation of elements is zero),

$$\Delta_r H^0 = \Delta_f H^0[CO(g)] - \Delta_f H^0[H_2O(g)]$$

and, from tables (Appendix 1), noting that the standard heat of formation of water in its gaseous state is different from that in its liquid form,

$$\Delta_r H^0 = -110.53 - (-241.82) = 131.29 \, \text{kJ mol}^{-1}.$$

We find the reaction is endothermic (it absorbs heat) and heat must be supplied to sustain it. The standard enthalpies of formation are tabulated at 298 K, whereas, in practice, the reaction is carried out at much higher temperatures. We can calculate the enthalpy of the reaction at a higher temperature, say 998 K, by using Kirchhoff's equation (Section 2.7). If we assume that the heat capacities of the reactants and products are independent of temperature, we can write

$$\Delta_r H_2 = \Delta_r H_1 + \Delta_r C_P (T_2 - T_1).$$

The heat capacities of compounds and the elements are given in Appendix 1.

$$\Delta_r C_P = C_P(H_2) + C_P(CO) - C_P(C) - C_P(H_2O)$$
$$\Delta_r C_P = 28.82 + 29.14 - 8.53 - 33.58 = 15.85 \, J \, K^{-1} \, mol^{-1}$$

and

$$\Delta_r H_{998} = \Delta_r H_{298} + 15.85(700)10^{-3} = 131.29 + 11.10 = 142.39 \, kJ \, mol^{-1}.$$

The reaction is somewhat more endothermic at high temperatures. The variation of the heat capacities with temperature can also be taken into account if the data are available. An accurate value of the enthalpy change accompanying the reaction is important because the temperature of the charcoal or coke employed can be maintained by passing air over it and using the exothermic reaction of oxygen and carbon

$$C(s) + \frac{1}{2}O_2(g) \rightarrow CO(g) \quad \Delta_f H^0(CO) = -110.53 \, kJ \, mol^{-1}$$

to provide the necessary heat. From a calculation of the enthalpy change accompanying this reaction at 998 K, the appropriate quantities of air and steam can be calculated in order to maintain the process in a steady state. The enthalpies of formation of selected substances can also be used to estimate the energies of individual bonds which, though they are not strictly constant from one molecule to another, can provide a useful guide to the enthalpy of formation of other compounds.

2.9 Calorimetry

The determination of thermodynamic quantities such as enthalpies of formation and heat capacities is carried out in *calorimeters*. Calorimetry allows the substance under investigation to be thermally isolated and any temperature changes in the system to be matched by a known quantity of electrical heating. The apparatus required to make these measurements depends on the state of matter of the substances involved.

When attempting to determine, for example, the *enthalpy of formation of methane*,

$$C(s) + 2H_2(g) \rightarrow CH_4(g),$$

we find that it is not easy to carry out the direct reaction under conditions in which the enthalpy change can be measured. However, it is relatively easy to measure the enthalpy changes accompanying combustion of hydrogen and methane using a *flame calorimeter*. In this apparatus, the gas flows into a combustion chamber along with oxygen and is burnt. The chamber is immersed in water and the temperature rise generated by the reaction is matched by a known quantity of electrical heating. The heat of combustion of solids can be measured in an apparatus called a *bomb calorimeter*. A measured quantity of the solid is ignited in an atmosphere of excess oxygen contained in a sealed vessel immersed in a quantity of water. The rise in temperature of the water due to the combustion is noted. A similar temperature rise is generated by a known amount of electrical heating and the comparison enables the heat of combustion to be determined.

We obtain from the results of these combustion experiments,

$$\Delta H/\text{kJ mol}^{-1}$$

$$C(s) + O_2(g) \rightarrow CO_2(g) \qquad\qquad -393.5$$

$$2H_2(g) + O_2(g) \rightarrow 2H_2O(l) \qquad\qquad -571.7$$

$$CO_2(g) + 2H_2O(l) \rightarrow CH_4(g) + 2O_2(g) \quad 890.4$$

(The values are for reactions in which the number of moles of the substances are specified by their stoichiometric coefficients.)

Adding these reactions gives us the result

$$2H_2(g) + C(s) \rightarrow CH_4(g) \qquad -74.8\,\text{kJ mol}^{-1}$$

Many determinations of this type have enabled the data given in Appendix 1 to be compiled.

Heat capacities of solids and liquids over a wide range of temperature can be determined in an apparatus known as an *adiabatic vacuum calorimeter* (Fig. 2.6). This allows a known amount of electrical energy to be used to heat up a sample in a

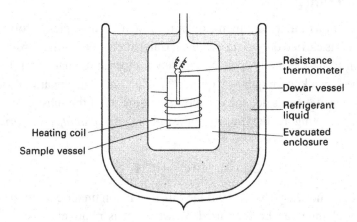

Fig. 2.6. Adiabatic vacuum calorimeter.

vessel thermally isolated by a vacuum. The temperature rise is measured, enabling the heat capacity to be directly calculated from the relationship $C = (dE/dT)$.

2.10 Coulombic energy

The most important source of potential energy in chemical systems is electrical. Two electrical charges, q_1 and q_2, a distance r apart, interact in a vacuum according to Coulomb's law to produce a potential energy, $V(r)$,

$$V(r) = \frac{q_1 q_2}{4\pi\varepsilon_0 r}.$$

The constant ε_0 is called the permittivity of free space. $4\pi\varepsilon_0$ has the value 1.11×10^{-10} J^{-1} C^2 m^{-1}. Each charge produces a *coulombic potential*, $\Phi(r)$, which falls off with distance according to the equation

$$\Phi(r) = \frac{q}{4\pi\varepsilon_0 r}.$$

The coulombic potential, $\Phi(r)$, is the potential energy that would be possessed by a unit charge situated at a distance r from a charge q. The magnitude of the electrostatic energy of two particles with the charge of an electron or proton a distance of 10^{-10} m apart is $V = 2.3 \times 10^{-18}$ J which, for a mole of such charges, would be 1.4×10^3 kJ - an energy comparable to that of the strongest chemical bonds. When two charges are immersed in a substance (other than a vacuum), the potential energy becomes

$$V = \frac{q_1 q_2}{4\pi\varepsilon_0 \varepsilon r},$$

where ε is the *relative permittivity* or *dielectric constant* of the substance separating the charges. The dielectric constant, which is greater than 1, has the effect of reducing the energy and interaction of electrical charges. Substances with a very high dielectric constant, such as water, for which it is approximately 80 at room temperature, can greatly reduce the energy of interaction. It is this reduction that facilitates the ionization of substances when dissolved in water.

Most molecules do not possess an overall charge and yet they still can give rise to an electrostatic potential energy if they contain separated charges. Consider a molecule containing two charges, q and $-q$, separated by a distance x. In a uniform electrostatic potential, the energies will cancel out to produce no net effect. However, if the electrostatic potential has a gradient $(d\Phi/dr)$, the potential energy experienced

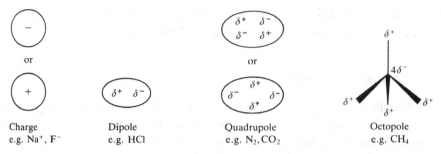

Fig. 2.7. Schematic representation of electric multipoles: the examples illustrated indicate only the lowest nonzero moment for each case.

by each charge is different.

$$V = q\Phi - q\left[\Phi + \left(\frac{d\Phi}{dr}\right)x\right] = -qx\left(\frac{d\Phi}{dr}\right)$$

The charge times the separation, qx, is defined as the *dipole moment*, μ, of the array of charges (Fig. 2.7). The gradient of the electrostatic potential is referred to as the *electrostatic field*, F. Thus, for a molecule which is overall electrically neutral but for which the centres of positive and negative charge do not coincide, the electrostatic potential energy can be expressed $V = -\mu F$. A convenient unit which is often used to express the dipole moment of molecules is the debye, D, where $1\,D = 3.34 \times 10^{-30}\,C\,m$. The dipole moment of two electronic charges separated by 10^{-10} m is 4.8 D. The dipole moment of symmetrical molecules is zero but that of water, H_2O, which is non-linear, is 1.8 D.

Molecules which possess neither an overall electrical charge nor a dipole moment, such as a linear molecule like nitrogen or carbon dioxide, may also interact through coulomb forces (Fig. 2.7). Such molecules possess a *quadrupole moment*, Θ, which is defined as the sum of the charges multiplied by the square of the distance from an appropriate origin. Molecules with quadrupole moments do not interact with the electrical potential or the field, but with the field gradient, (dF/dr), which is the second-order derivative of the electrostatic potential, $(d^2\Phi/dr^2)$. If a charge distribution is exposed to an external electrostatic potential, as might be generated by another similar charge distribution, we can express the resultant energy (considering all the charges to lie on only one axis, the z axis) as

$$V = q\Phi - \mu\left(\frac{d\Phi}{dz}\right) + \frac{1}{2}\Theta\left(\frac{d^2\Phi}{dz^2}\right).$$

The total charge, q, interacts with the electrostatic potential, the dipole moment, μ, with the field and the quadrupole moment, Θ, with the field gradient. We will return to the interaction of charge distributions when we consider the electrostatic contributions to the forces between molecules in more detail in Chapter 9.

2.11 Summary of key principles

The properties of molecules are determined, to a large degree, by the energy they possess. The total energy of an isolated molecule is made up of a number of contributions, *electronic, vibrational, rotational* and *translational* which, to a good approximation, can be regarded as independent.

$$E = E_{elec} + E_{vib} + E_{rot} + E_{trans}.$$

In addition, in phases where the molecules can interact with each other, the intermolecular forces give rise to intermolecular potential energy. Energy can be transferred either as *heat* (that energy which flows across temperature gradients) or *work* (see definition in Section 2.4). The total energy gained or lost by a system is given by

$$\Delta U = q + w,$$

where q is the heat absorbed by the system and w is the work done *on* the system. For all changes, the total energy of the system remains constant — the *First Law of Thermodynamics*.

Most chemical systems (other than those doing electrical work) only do the work of expansion or contraction against the atmospheric pressure, $w = -P\Delta V$. At constant volume, $w = 0$ and $\Delta U = q_V$, the heat absorbed by the system at constant volume. However, most chemical processes are carried out at constant pressure and $\Delta U = q - P\,dV$. We define enthalpy, $\Delta H = \Delta U + P\Delta V$, which is equal to the heat absorbed at constant pressure, q_P. U and H are state functions whose values depend only on the state of the system. This allows the enthalpy change, ΔH, for reactions that are difficult to measure directly to be determined from a series of more experimentally-accessible reactions which, when taken together, are equivalent to the reaction under consideration (Hess's law).

Problems

(1) Calculate the root-mean-square speed of chlorine molecules at 273 K. The molar mass of chlorine molecules is 70.9 g mol^{-1}.

(2) The pressure of an unknown gas effusing through a small orifice from a container drops to half its original value in 60 s. The pressure of nitrogen gas effusing under the same conditions drops to half its original value in 48 s. Estimate the relative molecular mass of the unknown gas. (See Section 2.2.)

(3) Calculate the work done by a gas when one mole expands from a volume of 1 dm^3 to 10 dm^3 (a) against a constant pressure of 1 atm, and (b) if the gas expands at a constant temperature of 298 K. (1 atm $= 1.01 \times 10^5$ Pa)

(4) The standard enthalpy of formation of ethene (C_2H_4) at 298 K is 52.3 kJ mol^{-1}. Estimate the enthalpy of formation of this substance at 500 K. At 298 K, the heat capacity of carbon is 8.53 J K^{-1} mol^{-1}, of hydrogen 28.8 J K^{-1} mol^{-1}, and that of ethene 43.5 J K^{-1} mol^{-1}.

(5) Calculate the standard enthalpy of formation of ethane (C_2H_6) at 298 K. At this temperature, the enthalpy of combustion of ethane is -1560 kJ mol^{-1}, that of carbon (graphite), -393.5 kJ mol^{-1} and that of hydrogen gas (H_2), -285.8 kJ mol^{-1}.

(6) Calculate the energy of interaction of the ions Na$^+$ and Cl$^-$ in a vacuum at a separation of 200 pm (relative to the energy at infinite separation). What would be the energy of interaction in water? (Take the relative permittivity (dielectric constant) of water as 80.)

3

The First Principle: Energy is Not Continuous

Towards the end of the 19th century, physicists felt that they had achieved enormous success in their attempts to understand the physical universe. The behaviour of mechanical objects and celestial bodies could be explained by Newtonian physics and electromagnetic radiation by the theories of James Clerk Maxwell. True, they recognised that a few loose ends remained to be tidied up but they were confident that these would be explained by a subtle application of the laws of what we now refer to as *classical physics*. They were quite wrong. The 'loose ends' proved to be the heart of modern physics and new and original ideas were required before they were to be understood.

3.1 The failures of classical physics

Some of the important experimental observations that classical physics was unable to explain are:

(i) Electromagnetic radiation can interact with matter and, if certain conditions are met, the radiation can be absorbed or emitted. To be able to understand and interpret the resulting, often complex, patterns of absorption or emission *spectra*, we need an understanding of the behaviour of atoms and molecules that cannot be provided by classical physics.

(ii) The colours of hot objects were also hard to interpret. It was well known that when objects were heated they became progressively red, yellow and finally white or even blue (Fig. 3.1). Classical physics, however, predicted that objects should emit most radiation in the blue and ultraviolet regions of the spectrum and that the total energy emitted over the whole of the spectral range should be infinite! This paradox was called the *ultraviolet catastrophe*.

(iii) The principle of equipartition of energy was found not to hold at low temperatures. In particular, the *heat capacity of ionic and atomic solids* approached zero at low temperatures and did not follow Dulong and Petit's law, which suggested they should remain constant at the value of $3R$ at all

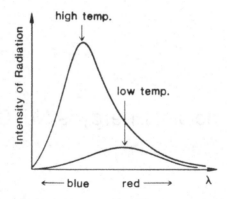

Fig. 3.1. Colour of hot objects (black body radiation).

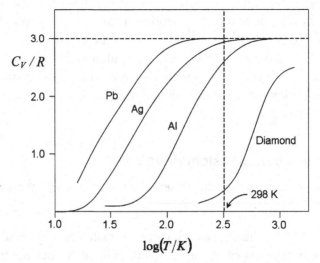

Fig. 3.2. The heat capacities of solids as a function of temperature: the heat capacities tend to zero at the absolute zero of temperature and to $3R$ at high temperatures (as predicted by the Dulong and Petit law). Solids with strong internal forces, such as diamond, tend to reach the high temperature limit only at very high temperatures.

temperatures (as predicted by the equipartition rule of classical physics for atoms that vibrated in three dimensions about their lattice points) (Fig. 3.2).

(iv) Another problem which defied explanation was the *photoelectric effect*. When certain metals were exposed to light, they could emit electrons. However, it was found that, if the frequency of light was below a critical value, no electrons were expelled regardless of the intensity of the radiation.

The search for solutions to these and other difficulties led to totally new laws of physics which were to revolutionise our view of the physical world and provide the basis for an understanding of chemical phenomena. The new theory that provided this understanding is called *quantum mechanics* or *wave mechanics*. A most important difference between the new and the old physics is the recognition that energy is not continuous.

3.2 Basic ideas of quantum mechanics

It was found that many of the failures of classical physics described above could be explained if the following bold assumption was made: *energy comes in discrete packets and is not continuous.* Max Planck, in 1900, first introduced this radical new idea, *quantisation*, to explain the colour of hot objects (black-body radiation). Albert Einstein extended its application to electromagnetic radiation by suggesting that the radiation itself consisted of small packets of energy, *photons*, with energy related to their frequency, v, by

$$E = hv,$$

where h is Planck's constant, 6.626×10^{-34} J s. Red light consists of low energy photons whereas blue or ultraviolet radiation is comprised of photons of higher energy. This idea provides a direct explanation of the photoelectronic effect. If the radiation is of too low a frequency, each photon has insufficient energy to dislodge electrons from the metal. A critical minimum energy, and a related frequency, v_c, given by $E_c = hv_c$, is necessary.

The photons that comprise electromagnetic radiation in the blue or ultraviolet part of the electromagnetic spectrum correspond to radiation with a frequency on the order of magnitude 10^{15} s^{-1}. Their energy is

$$E = hv = 6.626 \times 10^{-34} \text{ J s } \times 10^{15} \text{ s}^{-1}.$$

The energy transferred if a mole of photons of this frequency is absorbed by a material is

$$E = 6.626 \times 10^{-19} \text{ J } \times 6.022 \times 10^{23} \text{ mol}^{-1} = 399 \text{ kJ mol}^{-1}.$$

Electromagnetic radiation of this frequency can be absorbed by electrons in atoms or molecules and the energy absorbed is comparable with the strength of many chemical bonds. Ultraviolet light is often used to dissociate molecules and initiate chemical reactions.

The new assumptions provided an explanation of the spectra of atoms. If the atoms can exist in only a limited number of states with definite energy, then the energy differences between these states will also have only discrete values. For radiation to be absorbed, the energy of the photons must correspond to these energy differences, as

$$\Delta E = h\nu.$$

Therefore, the spectra consist of lines of specific frequencies. It also proved possible to use the fact that energy is quantised to explain the colours assumed by hot bodies and to show why the ultraviolet catastrophe is not observed. In fact, this was the first problem to be solved using the ideas of quantum mechanics by Max Planck (in his 1900 paper) that laid the foundation of modern physics and chemistry.

An experiment by Davisson and Germer in 1927 gave further insight into the new physics (Fig. 3.3). They observed that, if electrons were accelerated by oppositely charged plates and scattered by a crystal of nickel, a diffraction pattern was observed. Similar patterns can be produced with X-rays which are high-energy electromagnetic radiation. They had discovered that the beam of particles, the electrons, had wave-like properties. This lack of distinction between waves and particles had been predicted in 1924 by the French physicist Louis de Broglie. De Broglie noted that the energy of photons could be expressed as $E = h\nu$ and also, following the special theory of relativity, as $E = mc^2$, where c is the speed of light. Equating these results, and using the relation $\nu = c/\lambda$, gave

$$mc^2 = h\nu = \frac{hc}{\lambda}$$

and the momentum of the photon could be written

$$mc = \frac{h}{\lambda},$$

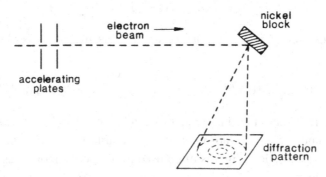

Fig. 3.3. The observation of interference in beams of electrons: the diffraction pattern is caused by the interference of electrons scattered from different layers of metal atoms in the metal block. The calculated wavelength is in good agreement with the de Broglie relation $m\nu = h/\lambda$. This experiment was first performed by Davisson and Germer in 1927.

an equation originally derived by Einstein. (It may seem surprising that a particle of zero rest mass can be held to possess momentum. This is one of the many paradoxes that arise in quantum mechanics. Niels Bohr once remarked that if one is not shocked by quantum mechanics, one has not understood it!)

De Broglie argued that this result might apply not only to photons, but to all moving particles of mass m and moving with velocity v. The momentum, p, is equal to mv, allowing us to write

$$m\text{v} \equiv p = \frac{h}{\lambda}$$

The wavelengths of the electrons calculated from the diffraction patterns observed by Davisson and Germer were in excellent agreement with de Broglie's equation. The wavelength of electrons travelling at $1 \times 10^7\,\mathrm{m\,s^{-1}}$ is given by

$$\lambda = \frac{h}{m\text{v}} = \frac{6.6 \times 10^{-34}\,\mathrm{J\,s}}{(9.1 \times 10^{-31}\,\mathrm{kg} \times 1 \times 10^7\,\mathrm{m\,s^{-1}})}$$

$$= 7.3 \times 10^{-11}\,\mathrm{m} = 70\,\mathrm{pm}\ (0.7\,\text{Å}) \quad \text{(Note: } \mathrm{J} = \mathrm{kg\,m^2\,s^{-2}}\text{)}$$

The wavelength is of the same order of magnitude as atomic dimensions, 100 pm. However, if we calculate the wavelength of an 'everyday' particle such as a tennis ball (mass ~50 g) which, if hit hard, travels at about 90 miles per hour (approximately $40\,\mathrm{m\,s^{-1}}$), we find

$$p = m\text{v} = 0.05\,\mathrm{kg} \times 40\,\mathrm{m\,s^{-1}}$$

$$\lambda = \frac{h}{p} = \frac{6.6 \times 10^{-34}\,\mathrm{J\,s}}{2.0\,\mathrm{kg\,m\,s^{-1}}} \approx 10^{-34}\,\mathrm{m},$$

a length so very small that it is impossible to detect the effects of the wave character of the tennis ball.

This fact that waves and particles are not different phenomena but can both be associated with the same objects is called *wave-particle duality*. It lies at the heart of modern science. The key to understanding this duality was provided by Max Born (1926). He realised that the amplitude of the wave at any point is related to the chance of finding the particle at that point. In fact, this probability is directly proportional to the square of the amplitude. A wave moving forward represents a beam of particles located where the amplitudes are largest. We will find this idea easier to understand when we consider particles that are confined to a small region of space. We can use the de Broglie equation to calculate the values of the energy that are allowed for such a particle and to examine how its behaviour can be expressed in terms of waves.

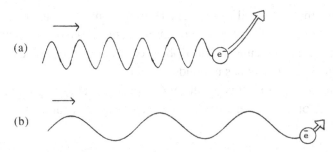

Fig. 3.4. The uncertainty principle: the wavelength of the photon in (a) is short and the position of the electron can be defined accurately. However, the high energy of the photon imparts a high, but unknown, velocity to the electron. In case (b), the energy of the photon is low and the electron does not acquire a significant velocity. However, the wavelength is so long that the position of the electron cannot be accurately defined.

3.3 The uncertainty principle

A most interesting consequence of wave-particle duality is the uncertainty principle first proposed by Werner Heisenberg in 1927. The principle states that the *exact position and momentum of a particle cannot be known simultaneously.* He had reached this conclusion by considering hypothetical experiments in which photons are scattered off a particle (so that it could be 'seen') (Fig. 3.4). The limit of resolution that can be obtained from the scattering of waves is on the order of their wavelength. If the photons are of very short wavelength, they can define the position with high accuracy, but because of their high energy ($E = h\nu = hc/\lambda$) the particle will be knocked away and its velocity will be uncertain. On the other hand, low-energy light with a long wavelength cannot define the position of the particle accurately. If we know the position precisely then we will be very uncertain of the velocity, and vice versa. Heisenberg expressed this principle in terms of the relation

$$\delta x \delta p \geq \frac{h}{4\pi},$$

where δx and δp are the uncertainties (standard deviations) in position and momentum respectively. For the particles we deal with in the everyday world, this uncertainty is not detectable, but for atomic particles the uncertainty principle sets important limits on what we can know and introduces non-deterministic features which are quite different from the laws of classical physics. One very important consequence of the uncertainty principle is that if we confine a particle to a restricted space, so that δx is small, its energy must be uncertain and it cannot have zero energy. The energy that results from this restriction is called the *zero-point energy*. Zero-point energy plays an important part in determining the behaviour of atomic and molecular systems.

We can express the uncertainty principle in another form involving the uncertainties in an energy of a state and its lifetime. Since $p = mv$ and $E = mv^2/2$, we have $E = p^2/2m$. Differentiating, we obtain

$$dE = \left(\frac{p}{m}\right) dp \quad \text{and} \quad dp = \left(\frac{m}{p}\right) dE.$$

Since $p/m = v = \delta x/\delta t$, we obtain $dx = (p/m)dt$ and, substituting dx and dp for δx and δp in the equation for the uncertainty principle given above, we obtain

$$\delta x \delta p = \left(\frac{p}{m}\right) \delta t \left(\frac{m}{p}\right) \delta E \geq \frac{h}{4\pi} \quad \text{and}$$

$$\delta E \delta t \geq \frac{h}{4\pi}.$$

Expressing the uncertainty principle in this way, in terms of energy and time, tells us that the accuracy with which the energy of a state can be determined is limited by the time over which it can be observed. *The shorter the lifetime of a state, the more uncertain is its energy.* Conversely, stable states have energies that can be precisely determined.

3.4   Summary of important principles

The energy of photons is related to the frequency of the radiation they transmit by

$$E = h\nu.$$

Matter and radiation can exhibit the properties of both particles and waves. This is called *wave-particle duality*. The wavelength of a particle is linked to its momentum by the *de Broglie relationship*,

$$m\nu = \frac{h}{\lambda}.$$

The square of the magnitude of the amplitude of the wave at any point in space is proportional to the probability of finding a particle at that point.

A consequence of wave-particle duality is that the momentum and position of a particle cannot simultaneously be precisely defined. The uncertainties in position and momentum are defined by *Heisenberg's uncertainty principle*,

$$\delta x \delta p \geq \frac{h}{4\pi}.$$

A further consequence that arises from this uncertainty is that a particle confined to a restricted space cannot have zero energy. The energy that results from this restriction is called the *zero-point energy*.

The application of these principles to chemical systems is the key to understanding how these systems behave. Gaining this understanding is the task we now have to undertake.

3.5 Translational motion: Particle in a box

To understand some of the implications of the principles set out above for small particles, for example electrons, atoms or molecules, let us imagine a particle of mass m confined to a 'box'. We will consider the box to be of only one dimension (the three-dimensional case is not much more difficult) and of length a. The potential energy of the particle outside the confines of the box is infinite and, within the box, in the region $0 < x < a$, it is zero. Let us assume that, since the chance of finding the particle is zero at the walls of the box, the amplitude of the associated wave must also be zero at the boundaries. This condition is satisfied if

$$\lambda = \frac{2a}{n},$$

where $n = 1, 2, 3$, etc. (Fig. 3.5). Using the de Broglie equation

$$\lambda = \frac{h}{m\mathrm{v}},$$

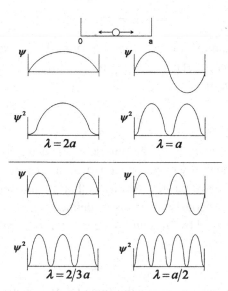

Fig. 3.5. Particle in a box: the wave functions, ψ, and the square of the wave functions, ψ^2 (which determine the probability of finding a particle in an element of space) for a particle confined in a one-dimensional box.

we obtain

$$\frac{2a}{n} = \frac{h}{mv}, \quad v = \frac{nh}{2ma}.$$

The particle has no potential energy within the box so that its total energy is kinetic:

$$E = \frac{m(nh/2ma)^2}{2}$$

$$E = \frac{n^2h^2}{8ma^2}$$

The particle can have only those energies for which n is a positive integer (Fig. 3.6). n is called a *quantum number*.

The lowest energy that the particle can have corresponds to $n = 1$. The particle cannot have zero energy because, since the particle is confined to a limited region of space, and δx is finite, this would contravene the uncertainty principle. The lowest energy level has the energy $E = h^2/8ma^2 = v^2/2$. Thus $p^2 = 2mE = h^2/4a^2$ and $p = \pm h/2a$.

As the momentum can take either value with equal probability, we can regard the uncertainty in momentum as the difference between these values, so that $\delta p = h/a$. The uncertainty in position is the length of the box, a, so that $\delta x \delta p = ah/a = h$, a value consistent with the uncertainty principle. This supports the view that zero-point energies arise from the uncertainty principle. We see that, for particles of low mass confined to small regions of space, the allowed energy levels become widely separated.

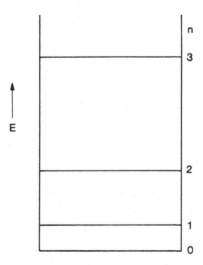

Fig. 3.6. The energy levels of a particle in a box.

The equation that represents the waves for a particle in a box is

$$\psi(x) = \psi^{max} \sin\left[n\left(\frac{\pi}{a}\right)x\right].$$

$\psi(x)$, the amplitude of the wave, is called the *wave function* of the particle and its square, $\psi^2(x)$, is proportional to the chance of finding the particle at x. (When ψ is a complex quantity, and has an imaginary component, $\psi^2(x)$ should be expressed as $\psi\psi^*$, where ψ^* is the *complex conjugate* of ψ. In the complex conjugate, the imaginary term $i = \sqrt{-1}$ is replaced by $-i$; $\psi\psi^*$ is always real.) For the probability of finding a particle to be exactly equal to $\psi^2(x)$, we must ensure that the total probability of finding it over all space is unity. To do this, we need to add a factor (called a *normalisation factor*) to the wave function to ensure that when integrated over all space, $\int \psi^2(x)d\tau = 1$.

Let us examine two situations of interest to chemists in which particles are confined to a limited region of space.

First: *The energy levels of an electron confined to a space of atomic dimensions.* Consider a one-dimensional box of atomic size for which the length $a = 10^{-10}$ m (100 pm).

$$E = \frac{n^2 h^2}{8ma^2}$$

$$= \frac{n^2(6.6 \times 10^{-34} \text{ J s})^2}{8 \times 9.1 \times 10^{-31} \text{ kg} (10^{-10} \text{ m})^2}$$

$$= n^2 \times 6.0 \times 10^{-18} \text{ J}.$$

For a mole of electrons in such a box, with $N_A = 6.0 \times 10^{23}$ mol^{-1}, the lowest level is

$$E = 6.0 \times 10^{23} \text{ mol}^{-1} \times 6 \times 10^{-18} \text{ J} = 3.6 \times 10^3 \text{ kJ mol}^{-1}.$$

The energy levels are far apart and would give rise to the absorption of electromagnetic radiation in the ultraviolet region.

Second: *The energy levels of a molecule confined to a box of 'laboratory' dimensions.*

For a nitrogen molecule confined to a one-dimensional box of a$=0.1$ m, the energy levels are given by

$$\frac{E}{n^2} = \frac{h^2}{8ma^2}.$$

Taking the atomic mass unit as 1.7×10^{-27} kg, we obtain, for the lowest energy level $(n = 1)$,

$$E = \frac{(6.6 \times 10^{-34} \, \text{J s})^2}{8 \times 28 \times 1.7 \times 10^{-27} \, \text{kg} \times (0.1 \, \text{m})^2} = 1.1 \times 10^{-40} \, \text{J}.$$

For one mole of nitrogen molecules, we obtain

$$E - 6.0 \times 10^{23} \, \text{mol}^{-1} \times 1.1 \times 10^{-40} \, \text{J} = 7.0 \times 10^{-17} \, \text{J mol}^{-1}.$$

The energy levels are so close together here that, for all practical purposes, we can regard the kinetic energy of the molecules as continuously distributed. Under these circumstances, the energy per degree of freedom (in this case translational motion in one dimension) will be the classical value, $RT/2$. At room temperature (298 K), $kT = 4.1 \times 10^{-21}$ J and, for one mole of substance,

$$RT = N_A kT = 2.5 \, \text{kJ mol}^{-1}.$$

Though our calculation was only for one-dimensional cases, we have obtained two important results which are quite generally true. First, the energy of electrons confined to spaces of molecular dimensions is highly quantised; that is, the allowed energies are very far apart. Second, the kinetic energy arising from molecules in what we could call laboratory-sized containers is essentially continuous. These conclusions are in keeping with the equation that tells us that the lower the mass of the particle and the smaller the space to which it has access, the further apart are the energy levels.

It is also instructive to consider the energy levels of a particle in a three-dimensional box. In this case, the energy levels are determined by three quantum numbers n_x, n_y and n_z. If the sides of the box are of lengths a_x, a_y and a_z in the x, y and z direction, the energy levels are given by

$$E = \left(\frac{h^2}{8m} \right) \left[\frac{n_x^2}{a_x^2} + \frac{n_y^2}{a_y^2} + \frac{n_z^2}{a_z^2} \right].$$

If the box is cubic, $a_x = a_y = a_z = a$, and then the energy levels are given by

$$E = \left(\frac{h^2}{8ma^2} \right) [n_x^2 + n_y^2 + n_z^2].$$

The lowest level corresponds to $n_x = 1$, $n_y = 1$, $n_z = 1$ and the sum of the values of the quantum numbers in the brackets is three. This $(1, 1, 1)$ is the only contribution to the lowest energy level which corresponds to one energy state. However, the next level, for which the sum of the numbers in the brackets is six, can be achieved in three different ways: $(1, 1, 2)$, $(1, 2, 1)$ and $(2, 1, 1)$. This level is said to be threefold *degenerate. The degeneracy of an energy level is the number of separate quantum states that have the same energy.* Because of the various possible combinations of three integers, the successive energy levels have degeneracies of $1, 3, 3, 1, 6, 3, \ldots$ and so on.

If we make a small distortion of the cube so that the dimensions become slightly different, we will see that most of the energy levels split into a number of components. This illustrates an important property of quantum mechanical systems — that degeneracies arise from the symmetry of the system. *When the symmetry is removed, so are the degeneracies.* This example provides insight into the behaviour of more complicated systems of greater interest to chemists which we will meet later.

3.6 Rotational motion

The energy of particles constrained in rotational motion, such as a particle following a circular orbit, is also quantised. The waves representing the circular motion must 'join up' if they are to be stable and not destroyed by mutual interference. The restriction that arises from this condition is that the circumference of a circle must be a unit number of wavelengths; that is, $2\pi r = n\lambda$ or $\lambda = 2\pi r/n$ (Fig. 3.7). If this condition is met, the waves become 'standing waves'. If the condition is not met, the waves will interfere destructively with each other. The de Broglie relationship gives $mv = h/\lambda$ and $\lambda = h/mv$. Combining this equation with the restrictive condition for wavelength, $\lambda = 2\pi r/n$, we obtain

$$\frac{2\pi r}{n} = \frac{h}{mv} \quad \text{and} \quad v = \frac{nh}{2\pi mr}.$$

The only energy associated with the rotational motion is kinetic energy given by

$$E = \frac{mv^2}{2} = \frac{h^2 n^2}{8\pi^2 mr^2} = \frac{h^2 n^2}{8\pi^2 I},$$

where the moment of inertia, $I = mr^2$.

Again, as for the particle in a box, we find that the energy levels are proportional to n^2, where $n = 0, 1, 2, \ldots$ is now the rotational quantum number (Fig. 3.8a). A subtle difference between the particle in a box and the rotational motion considered here is the fact that the value $n = 0$ is permitted because, when the rotating particle possesses zero angular momentum, its position (orientation) is unspecified and this is therefore consistent with the uncertainty principle.

The rotational motion we have just considered is that of circular motion in two dimensions. If we allow for the rotating mass to move in three dimensions, that is, over the surface of a sphere, rather than in a flat two-dimensional circle, we obtain

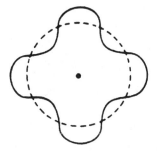 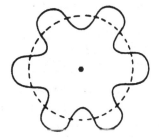

Allowed continuous waves.

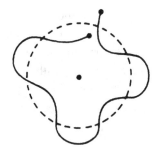

Waves do not join up.
This configuration is not allowed.

Fig. 3.7. The wave theory model of rotational motion.

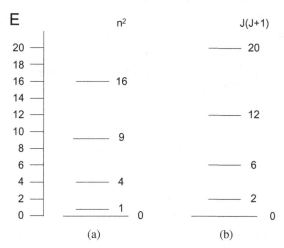

Fig. 3.8. Rotational energy levels of a rigid rotor: (a) two-dimensional rotation, $E \propto n^2$; (b) three-dimensional rotation, $E \propto J(J + 1)$.

a slightly more complicated expression for the energy levels

$$E = \frac{h^2 J(J+1)}{8\pi^2 I}, \quad \text{where } J = 0, 1, 2, \ldots,$$

and J is now the rotational quantum number. The energy levels are degenerate with a degeneracy of $(2J + 1)$ (Fig. 3.8b).

The same expression is obtained for a rigid dumbbell shape, two masses, m_1 and m_2, connected by a rigid link of length R. The moment of inertia in this case is given by

$$I = \left[\frac{m_1 m_2}{(m_1 + m_2)} \right] R^2 = \mu R^2.$$

$[m_1 m_2/(m_1+m_2)]$ is called the reduced mass, written μ (which must not be confused with the dipole moment, for which the same symbol is used).

This proves to be a satisfactory model for rotating diatomic molecules and enables us to calculate, to a reasonable degree of accuracy, the rotational energy levels of such molecules. Light diatomic molecules, such as hydrogen, with a small moment of inertia, have energy levels that are far apart when compared with heavier molecules such as the diatomic chlorine molecule. A similar, but slightly more complicated equation, defines the energy levels of three-dimensional polyatomic molecules in terms of the moments of inertia about the x, y and z axes.

3.7 Vibrational motion

The simplest and most common model of vibrational motion is a *simple harmonic oscillator*, which occurs when the restoring force is proportional to the distance from the position of equilibrium and the potential energy of the oscillator is proportional to the square of the amplitude (see Section 2.1).

The determination of the energy levels for motion governed by this force law involves quite difficult mathematics, but leads to a very simple result. The allowed energy levels are equally spaced

$$E = \left(v + \frac{1}{2} \right) h\nu,$$

where $v = 0, 1, 2, \ldots$ is the vibrational quantum number. The $\frac{1}{2}h\nu$ contribution to the energy when $n = 0$ is termed the *zero-point energy* of the oscillator, and is necessary for the solution to be consistent with the uncertainty principle.

The lower energy levels of vibrating molecules are usually almost equally spaced, showing that the molecular vibrations are well-represented by simple harmonic motion. However, as the vibrations become more energetic, the simple harmonic oscillator model fails and the higher energy levels become closer. Eventually, they converge, at which point the bond breaks and the molecule dissociates into two fragments.

3.8 The hydrogen-like atom

In the hydrogen-like atom, a single electron moves around a nucleus of charge Z. Z is equal to unity for hydrogen, but may take on other integer values for other hydrogen-like atoms and ions, characterised by a single electron in the presence of a charged nucleus, such as He^+, Li^{2+}, etc. We assume that it is necessary for a stable energy level that the electronic wave function is represented by a standing wave. The condition for this, as we discussed above, is that the circumference of the circular paths the electron can occupy are restricted by the equation $2\pi r = n\lambda$. For non-integral values of n, the waves will not 'join up'. Combining this with de Broglie's equation, written as

$$\lambda = \frac{h}{m\text{v}},$$

(where m is a mass of an electron and v its velocity), leads to

$$2\pi r = n\left(\frac{h}{m\text{v}}\right),$$

which can be written

$$(m\text{v})^2 = \frac{n^2 h^2}{4\pi^2 r^2}.$$

We combine this relation with the condition that the electrostatic force of attraction of the nucleus on the electron, $Ze^2/4\pi\varepsilon_0 r^2$, must equal the centripetal force on the electron, $m\text{v}^2/r$, so that

$$\frac{Ze^2}{4\pi\varepsilon_0 r^2} = \frac{m\text{v}^2}{r},$$

which gives

$$(m\text{v})^2 = \frac{Ze^2 m}{4\pi\varepsilon_0 r}.$$

Eliminating $(m\text{v})^2$, we obtain

$$r = \frac{n^2 h^2 4\pi\varepsilon_0}{Ze^2 4\pi^2 m},$$

which we can write as

$$r = \frac{n^2 a_0}{Z}, \quad \text{where } a_0 = \frac{(4\pi\varepsilon_0)h^2}{4\pi^2 m e^2} = 52.9\,\text{pm}.$$

a_0 is called the *Bohr radius* and is the most probable distance from the nucleus of an electron in the lowest energy state of a hydrogen atom for which $n = 1$ and $Z = 1$. The energy of the electron is given by

$$E = \text{potential energy} + \text{kinetic energy}$$

$$E = \frac{-Ze^2}{4\pi\varepsilon_0 r} + \frac{mv^2}{2}.$$

Substituting for r and $(mv)^2$, we obtain

$$E = -\frac{2\pi^2 m e^4 Z^2}{n^2 h^2 (4\pi\varepsilon_0)^2} = \frac{-R_H Z^2}{n^2}.$$

Evaluating the constants to obtain the *Rydberg constant*, R_H, gives

$$R_H = 1312\,\text{kJ mol}^{-1} = 109\,737\,\text{cm}^{-1} = 13.60\,\text{eV}.$$

Just as the Bohr radius was identified as an appropriate unit for atomic and molecular calculations, it is convenient to define a unit of energy, called the *hartree*, E_h, which is equal in value to twice the Rydberg constant, $2R_H$.

$E_h = 4\pi^2 m e^4 / h^2 (4\pi\varepsilon_0)^2$ and has the value $2625\,\text{kJ mol}^{-1}$, $27.21\,\text{eV}$ or $219474\,\text{cm}^{-1}$. The atomic units E_h and the Bohr radius, a_0, are very convenient to use when performing quantum mechanical calculations.

In the above theory, r, the distance of the electron from the nucleus is obtained as a precise number. However, the electrons cannot follow a specific orbit as this would not be consistent with the probabilistic nature of quantum mechanics, which requires that the trajectories of electrons cannot be precisely defined. Rather than use the word orbits, we define *atomic orbitals* which are related to the *probability* of finding electrons in any particular region of space. An atomic orbital is a wave function for a single electron. The probability of finding an electron in any region of space is proportional to the square of the wave function, ψ. For the ground state, i.e. the lowest energy level, of the hydrogen-like atom, the wave function is given by

$$\Psi = \left(\frac{1}{\pi}\right)^{1/2} \left(\frac{Z}{a_0}\right)^{3/2} \exp\left(\frac{-Zr}{a_0}\right)$$

and the probability of finding an electron in any region of space dV is given by

$$\rho = \left(\frac{1}{\pi}\right)\left(\frac{Z}{a_0}\right)^3 \exp\left(\frac{-2Zr}{a_0}\right) dV.$$

Although this function exhibits simple exponential decay, the probability of finding an electron at a distance r from the nucleus, as illustrated in Fig. 3.9, does not. This is because the volume of space at a distance r from the nucleus is given by $dV = 4\pi r^2 dr$ and the function $4\pi r^2 \rho$ goes through a maximum value at $r = a_0$, thus confirming that a_0 is the most probable distance of the electron from the nucleus.

The results that we have obtained for the energy levels and orbital sizes of hydrogen-like atoms were first obtained by Niels Bohr in 1913 using a different approach. He made the assumption that the angular momentum of an electron orbiting a nucleus was quantised in units of $h/2\pi$. This assumption led to the same results as those obtained above.

3.9 Hydrogen atom spectra

We will need to discuss the spectra of atoms in some detail in Chapter 6, but that of the hydrogen atom is particularly simple and the regularities we observe provide direct confirmation of the success of quantum mechanics. The electronic energy levels of the hydrogen atom are determined by the principal quantum number, n,

$$E = -\frac{2\pi^2 m e^4 Z^2}{n^2 h^2 (4\pi\varepsilon_0)^2} = \frac{-R_H Z^2}{n^2}.$$

The transitions between the energy levels a and b, with the corresponding quantum numbers n_a and n_b, given by this equation, are

$$\Delta E = R_H \left[\frac{1}{n_a^2} - \frac{1}{n_b^2}\right].$$

This results in a number of series of lines, each originating from a particular level defined by n_a, with n_b taking the values $(n_a + 1), (n_a + 2), \ldots$ until the energy spacings converge on zero (Fig. 3.10).

When $n_a = 1$ and $n_b = 2, 3, \ldots$ the lines are called the Lyman series. The lines that arise from the $n_a = 2$ level are called the Balmer series. Other similar series originate from larger values of n_a and are termed the Paschen, Brackett, Pfund etc. series. For each series, the limit is reached when $n_b = \infty$ and $1/n_b^2 = 0$. This limit gives us the energy required to remove an electron from each of the levels defined by n_a, called the ionization energy, I, of the level, given by

$$I = \frac{R_H}{n_a^2}.$$

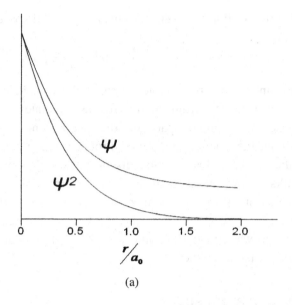

(a)

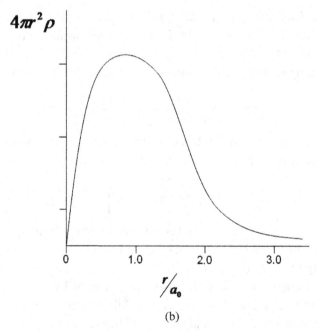

(b)

Fig. 3.9. (a) The wave function, ψ, and the square of the wave function, ψ^2 for the ground state of the hydrogen atom; (b) the probability of finding an electron, $4\pi r^2 \rho$, as a function of distance from the nucleus, r. ρ is the probability of finding the electron in a unit volume of space.

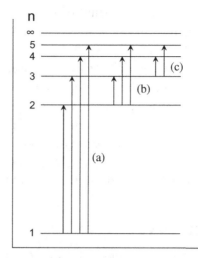

Fig. 3.10. The energy levels and spectral transitions of the hydrogen atom: (a) Lyman series; (b) Balmer series; (c) Paschen series. The energy levels for which $n > 5$ are not shown.

For the removal of an electron from the lowest energy level of the hydrogen atom, for which $n_a = 1$, we have

$$I_1 = 109737 \, \text{cm}^{-1} \quad (1312 \, \text{kJ mol}^{-1})$$

and, for $n_a = 2$,

$$I_2 = 27434 \, \text{cm}^{-1} \quad (328.1 \, \text{kJ mol}^{-1}).$$

We can apply these equations to systems, other than the hydrogen atom, where a single electron interacts with a charged nucleus. For example, we can calculate the first ionization energy of an electron in the lowest energy level energy level of Li^{2+}:

$$E = \frac{-R_H Z^2}{n^2}.$$

Since $n = 1$ and $Z = 3$ for lithium, we obtain $E = 987633 \, \text{cm}^{-1}$ ($11808 \, \text{kJ mol}^{-1}$).

We can also calculate the average distance of an electron from the nucleus in this energy level:

$$r = \frac{52.9 \, n^2}{Z} \, \text{pm}$$

$$r = \frac{52.9}{3} = 17.6 \, \text{pm}.$$

These relations that characterise the spectrum of hydrogen had first been noted in the 19th century. However, their theoretical basis was not understood until Bohr produced his theory in 1913. His expression for the Rydberg constant was essentially exact and his theory gave what appeared to be a complete account of the hydrogen atom. However, attempts to extend it to other atoms were not successful and it soon became apparent that hydrogen was a particularly simple case in which, apparently, spectroscopic transitions were allowed between all available energy levels. This simple picture arises from the fact that there is only one electron surrounding the nucleus so that the energy is determined only by the principal quantum number, n. This is not the case when more than one electron is involved.

3.10 The Schrödinger wave equation

The simple theory of hydrogen-like atoms works well for one-electron atoms and ions like He^+ or Li^{2+} but breaks down in other cases. The reason is that when an atom has more than one electron the interactions between the electrons lead to energy states that cannot be described in terms of n, the principal quantum number, alone.

Schrödinger proposed a modification of equations that govern wave motion which provided a way of applying the integral wavelength restriction, which we used above, in a more sophisticated manner. The Schrödinger equation is not simple, and solving it, even in the most straightforward cases, can involve difficult mathematics. Nevertheless, the equation is, without doubt, the most important and most-used approach to the quantum mechanical solution of physicochemical problems. It provides a way of calculating the *wave function* (which defines the probability of finding a particle in a particular region of space) and the *energy levels* of a particle from a knowledge of its kinetic and potential energies as a function of position. It is a differential equation which can be written for a one-dimensional system as

$$-\left(\frac{h^2}{8\pi^2 m}\right)\left(\frac{\mathrm{d}^2\psi}{\mathrm{d}x^2}\right) + V(x)\psi = E\psi,$$

where x defines the position of the particle, m its mass, E its total energy and $V(x)$ is the potential energy. For this equation, solutions will exist only for certain values of E, the allowed energy levels of the system, called *eigenvalues*. It can be shown that application of the Schrödinger equation leads directly to the energy levels of a particle in a box and those of the hydrogen-like atom which we obtained using de Broglie's relationship.

The equation is readily extended to three dimensions (using the partial differential coefficients introduced in Section 2.6):

$$-\left(\frac{h^2}{8\pi^2 m}\right)\left(\frac{\partial^2}{\partial x^2} + \frac{\partial^2}{\partial y^2} + \frac{\partial^2}{\partial z^2}\right)\psi + V(x, y, z)\psi = E\psi.$$

This can be written in a more compact form:

$$-\left(\frac{h^2}{8\pi^2 m}\right)(\nabla^2\psi) + V\psi = E\psi,$$

where $\nabla^2 = (\partial^2/\partial x^2 + \partial^2/\partial y^2 + \partial^2/\partial z^2)$ and E is a function of all three coordinates. ∇^2 is introduced as a new type of mathematical notation which we have not employed before. It is called an *operator* because it operates to change the functions that follow it. We can achieve an even more compact formulation of the Schrödinger equation if we define a further operator called the *Hamiltonian*, $\mathbf{H}$

$$\mathbf{H} = -\left(\frac{h^2}{8\pi^2 m}\right)\nabla^2 + V,$$

and the Schrodinger equation then becomes

$$\mathbf{H}\psi = E\psi.$$

If we multiply each side by ψ and integrate over all space, we obtain, for the energy,

$$E = \frac{\int \psi \mathbf{H}\psi\, d\tau}{\int \psi^2 d\tau}.$$

We must remember that $\mathbf{H}$ operates on, not multiplies, the ψ that follows it. (If ψ is a complex quantity then this equation becomes $E = \int \psi^*\mathbf{H}\psi d\tau / \int \psi^*\psi d\tau$; see Section 3.5.)

We should note that these equations are simply reformulations, for convenience, of the original Schrödinger equation. No new scientific principles have been added.

The integrals occurring in the above equations can be expressed more concisely in what is known as *bra-ket notation* first devised by Dirac. Thus, the integral over all space, $\int \psi_a^*\psi_b d\tau$, can be written $\langle a|b\rangle$, where $\langle a|$ is called a *bra* and $|b\rangle$ a *ket*. Integrals of the type $\int \psi_a^*\mathbf{H}\psi_b d\tau$ are expressed in this notation by $\langle a|\mathbf{H}|b\rangle$. It is important to be familiar with this notation. However, for maximum clarity, we will continue to write the integrals in full.

Unfortunately, exact solutions of the Schrödinger equation are too difficult to obtain for all but the simplest molecular problems. The factors determining the electrostatic potential energies of atoms and molecules are usually easy to formulate, so the exact expression for the potential energy (or, in the notation above, the exact Hamiltonian) is usually known. Unfortunately, even with this knowledge, it is not possible to solve the equation to obtain exactly the wave functions, ψ, and the allowed energies, E. However, a number of effective methods have been developed to provide approximate solutions of the Schrödinger equation and these are discussed in the next chapter.

3.11   Quantum mechanics — further considerations

We have introduced quantum mechanics through the concept of wave-particle duality identified by de Broglie and the wave equation proposed by Schrödinger. This approach provides working chemists with all the knowledge of the subject that is necessary for their day-to-day purposes. However, this is not the only route to quantum mechanics and, indeed, was preceded by an alternative and less accessible matrix algebra formulation due to Heisenberg. Paul Dirac developed yet another approach to quantum mechanics and, in 1928, showed that electron spin was a natural consequence of the theory of relativity. He used his new formulation to predict the existence of the positron and antimatter.

It is possible to develop quantum mechanical theory through a set of axioms and, though we will not attempt to pursue this approach in detail, it is instructive to consider some of the more general axioms. The first states that there exist continuous functions, referred to as state functions or wave functions, which fully determine what can be known about a system. Secondly, that each observable property is associated with an operator. The solutions of the equation in which the operator acts on the state function, called *eigenvalues*, correspond to the only values of the property that can be observed and measured. The operators replace the functions which describe physical properties in classical mechanics. Thus, the momentum, which in classical mechanics is described by the function $m(dx/dt)$, corresponds to the operator $(h/2\pi i)(d/dx)$ and the kinetic energy, classically described by $m(dx/dt)^2/2$, to the operator $-(h^2/8\pi^2 m)(d^2/dx^2)$. When two operators operate on a function and the order in which they operate is immaterial, the two operators are said to commute. However, in other cases, different results are obtained depending on the order in which the operators are applied. Under these latter circumstances, Heisenberg showed that it is impossible to measure precisely two properties simultaneously when their corresponding operators do not commute. Thus, momentum and position, and energy and time are examples where simultaneous precise measurements cannot be achieved, making it impossible to develop a deterministic theory of the physical world. The predictions of quantum mechanics are, at best, probabilistic.

Not all scientists were happy with this conclusion. Einstein, famously quoted as saying 'God does not play dice with the universe', believed that quantum mechanics provided only an interim theory and that, ultimately, a deterministic theory in terms of variables, now hidden from our view, would be developed. Einstein, Podolsky and Rosen, in 1935, formulated a paradox which stated that either (i) measurements on one part of the system could change the state of another remote part of the system or (ii) that quantum mechanics is incomplete. They felt that the improbability of (i) vindicated their view that the description given by quantum mechanics was an incomplete theory. By contrast, Bohr and his colleagues took the view, usually referred to as the Copenhagen interpretation, that quantum mechanics correctly describes a universe which is fundamentally nondeterministic and that the uncertainties arising from quantum mechanics are more than just a reflection of our limited knowledge. In 1964, John Bell proposed a theorem which stated that no theory with hidden variables could lead to all the predictions of quantum mechanics. He was able to suggest a number of experiments to test the theorem, which required the observation of the correlation of pairs of particles by distant, separated observers. Quantum mechanics predicts a greater degree of correlation than would be predicted by a theory with hidden variables. The results of these and other tests have been held to confirm the correctness of the Copenhagen interpretation and to rule out the existence of hidden variables. The universe does indeed appear to follow only probabilistic laws. Nevertheless, some optimistic physicists continue the search for a deterministic theory based on hidden variables.

Quantum mechanics has had an enormous impact on our whole understanding of the nature of scientific discovery. Prior to 1920, virtually all scientists would have assumed that the 'laws of nature' were deterministic in character. Following Laplace, they would have believed that a 'superior intelligence', furnished with all the knowledge of an existing state, could accurately predict its future behaviour. Following the development of quantum mechanics, it became impossible to hold this view with any confidence. The fundamental basis of classical physics was destroyed. The precision and certainty offered by scientific theory prior to the 20th century were replaced by probability and indeterminacy. This was, perhaps, the most important revolution in scientific thinking in 2000 years. Indeterminacy provides much for both scientists and philosophers to debate. It fuels wider speculation about the nature of free will and the role of individual responsibility, taking it far beyond the limited purposes of this book!

Problems

(1) Calculate the energy transferred to a substance if it absorbs 1 mole of photons of blue radiation of wavelength 4.0×10^{-7} m.

(2) Calculate the wavelength of a proton accelerated to a velocity of 1.0×10^5 m s^{-1}.

(3) After a particular electron has been accelerated, the uncertainty in its velocity is 5×10^4 m s^{-1}. With what accuracy can its position be defined?

(4) A proton is confined in a one-dimensional space of length 150 pm. Calculate the energy of its lowest energy level. What would be the energy of a mole of protons confined in this way?

(5) The first spectral lines of the Balmer series of the hydrogen atom are $(1.524, 2.058, 2.304$ and $2.438) \times 10^4$ cm^{-1}. Calculate the Rydberg constant and the series limit of the Balmer series.

4

Electrons in Atoms

4.1 Limitations of the simple model

In the last chapter, we found a quantum-mechanical solution to the energy levels of the hydrogen-like atom containing only one electron, which expressed the energy in terms of a single quantum number. Although essentially exact for the hydrogen atom, this solution concealed many complications that arise in the accurate solution of the Schrödinger equation for electrons moving in three dimensions about a nucleus. When we have more than one electron close to the nucleus, we find that we need three quantum numbers to give a proper description of the electronic energy levels. A fourth quantum number will be introduced later, in Section 4.3. A single electron moving around the nucleus is a system of perfect spherical symmetry and a simple model of 'joining up' the wave functions around the circumference of a circle leads to the correct energy levels. However, once we introduce another electron, the perfect spherical symmetry is broken and we find that the pattern of energy levels is more complicated and governed by three quantum numbers. This is consistent with the three-dimensional nature of the system. Although the energy levels of the hydrogen-like atom depend only on one quantum number, the energy levels are degenerate and this degeneracy is revealed when the perfect spherical symmetry is broken. When more than one electron is present, the interactions between the electrons cause a splitting of the energy levels. We consider this process in more detail below.

4.2 Solution of the Schrödinger equation for many-electron atoms

The three quantum numbers which define the wave functions that arise from the rigorous application of the Schrödinger equation to *hydrogen-like* atoms are:

(i) The *principal quantum number*, n, which we have already met. It is a positive integer and, for hydrogen-like atoms, totally determines the energy of an atomic energy level by $E_n = -R_{\mathrm{H}} Z^2 / n^2$, where R_{H} is the Rydberg constant and Z is the atomic number of the atom.

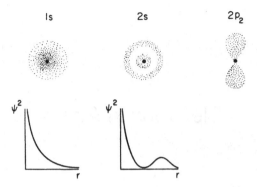

Fig. 4.1. The shapes of atomic orbitals: the density of shading represents the chance of finding an electron at a particular point.

(ii) The *(orbital) angular momentum quantum number, l*, determines the magnitude of the angular momentum of the electron which is given by $[l(l + 1)]^{1/2} h/2\pi$. *l* can take on any value in the range $0, 1, 2 \ldots (n - 1)$. For each value of *l*, the *spatial distribution* of the electrons follows a different pattern but, for hydrogen-like atoms, the energies are determined only by E_n. These distributions are called *orbitals*. The orbitals are the areas of space where ψ^2 is significant and there is a significant probability of finding the electron (Fig. 4.1). When $l = 0$, the orbitals have zero angular momentum, are spherically symmetrical and are called s orbitals. When $l = 1$, the orbitals have finite angular momentum, are no longer spherically symmetric, have two lobes and are designated p orbitals.

(iii) The *magnetic quantum number*, m_l, determines the direction in space in which the orbitals 'point' and determines the component of the angular momentum in a particular direction. The rule for m_l is that it can take the $(2l + 1)$ values

$$m_l = 0, \pm 1, \pm 2, \ldots, \pm l,$$

that is, any integer from $+l$ to $-l$. Thus, for s orbitals when $l = 0$, m_l can only be zero as the orbitals are spherically symmetric. However, for p orbitals, m_l can take on the values $-1, 0, +1$ and the orbitals can be oriented in one of three directions: p_x, p_y and p_z (Fig. 4.2). d orbitals, for which $l = 2$, can have five different orientations and, for f orbitals, where $l = 3$, there are seven orientations.

Thus, for a hydrogen-like atom, although the energy is determined solely by the principal quantum number, *n*, the energy levels are degenerate, with the *l* and m_l quantum numbers defining different quantum states with the same energy. The character and range of the atomic quantum numbers are summarised in Table 4.1.

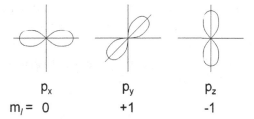

$$p_x \qquad p_y \qquad p_z$$
$$m_l = \ 0 \qquad\qquad +1 \qquad\qquad -1$$

Fig. 4.2. The directions of p orbitals: m_l can take three values that define which of the x, y and z axes the p orbitals are directed along. The labelling of the values of m_l is arbitrary.

Table 4.1. Atomic quantum numbers.

Quantum number	Name	Property determined	Range
n	Principal	energy	$1 \rightarrow \infty$
l	Angular momentum (azimuthal)	magnitude of orbital angular momentum, orbital shape	$0 \rightarrow (n-1)$
m_l	(Orbital) Magnetic	orbital direction	$+l \rightarrow 0 \rightarrow -l$
s	Spin	magnitude of electron spin angular momentum	$1/2$
m_s	Spin magnetic	direction of electron spin	$+1/2$ or $-1/2$

The analytical expressions for the wave functions that arise from the solution of the Schrödinger equation for hydrogen-like atoms are not simple. They are made up of two contributions, one radial, $R_{n,l}(r)$, and one angular, $Y_l^{m_l}(\theta, \varphi)$

$$\Psi_{n,l,ml}(r, \theta, \varphi) = R_{n,l}(r)Y_l^{m_l}(\theta, \varphi).$$

The radial terms are determined by the quantum numbers n and l and the angular terms (called spherical harmonics) by the quantum numbers l and m_l. Both terms have nodes where the wave function becomes zero. The nodes in the angular term are determined by the angular momentum and characterise the s, p, d etc. orbitals. The zeros in the radial term tend to occur close to the nucleus and are less important in determining chemical properties. Use is made of this fact when developing approximate wave functions that are convenient to use in calculations. The full expressions for s and p wave functions are given in Appendix 2.

4.3 Electron spin

The three quantum numbers derived from the solution of the Schrödinger equation for hydrogen-like atoms were very successful in explaining the behaviour of atoms. However, some phenomena, such as the fact that the strongest line in the sodium spectrum was made up of two closely-spaced lines, could not be understood. The explanation, put forward by George Uhlenbeck and Samuel Goudsmit in 1925, was that the electron behaves as if it were a spinning top with a magnetic moment $\mu_B = eh/4\pi m_e$, where m_e is the mass of an electron. Every electron has the same magnitude of spin angular momentum defined by the spin quantum number s. This can take only one value, namely 1/2. The magnitude of the spin angular momentum is given by $[s(s+1)]^{1/2}h/2\pi = (3/4)^{1/2}h/2\pi$. The electron spin can take only one of two orientations defined by another quantum number, the *spin magnetic quantum number*, m_s. One orientation corresponds to $m_s = +1/2$ and the other to $m_s = -1/2$.

As the spin of an odd, unpaired, electron can take on one of two values, small energy differences arise that can cause an atomic energy level to split into two closely-spaced levels to produce what is called a *doublet state*. As we will learn in Chapter 6, this can lead to a splitting of spectral lines, as occurs with the intense yellow line in the sodium spectrum. If two electrons have spins in opposite directions, the overall spin quantum number is zero and the state they give rise to is termed a *singlet state*. If two electrons have parallel spins, each having the same value of m_s, either $+1/2$ or $-1/2$, the total spin quantum number can take on the values of $+1$, 0, or -1 and energy levels can be split into three, leading to a *triplet state*. When atoms or molecules contain electrons with unpaired spins the substances they form are found to be *paramagnetic* and are attracted to a magnetic field.

4.4 Many-electron atoms

As the electrons in many-electron atoms can interact with each other, the wave functions depend on the coordinates of all the electrons. Such complexity is very difficult to handle and, to deal with many-electron atoms, we normally make a crucial approximation. We assume that each electron occupies its own orbital, defined by a hydrogen-like wave function. This approximation is the *orbital approximation*, which provides an effective basis for describing atomic structure and the behaviour of electrons in atoms. The energy of each electron depends on the effective nuclear charge it experiences which, in turn, depends on the ability of the other electrons in the atom to shield the nucleus. Electrons can prevent other electrons, particularly those lying further from the nucleus, from experiencing the full effect of the positive charge on the nucleus.

This enables us to understand why s and p orbitals with the same principal quantum number n have different energies. We can illustrate this by considering a case where outer electrons lie in orbitals beyond those of inner electrons, as illustrated in Fig. 4.3, where the open circles represent inner electrons that can shield the outer electrons from the attraction of the nucleus. We can see that the

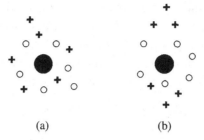

(a)   (b)

Fig. 4.3. Many electron atoms (schematic): the open circles represent the inner or core electrons. The crosses represent the outer or valence electrons. Case (a) *s* valence electrons and case (b) *p* valence electrons.

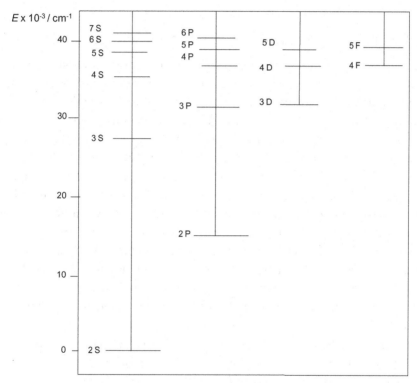

Fig. 4.4. The energy levels of the lithium atom. Higher levels not shown. (The spectroscopic notation is explained in Section 6.4.)

s electrons have a greater chance of lying inside the inner electrons than do the p electrons. Therefore, they experience more fully the attraction of a nucleus and have lower energy. This causes the levels with the same value of n to split, as is shown for lithium atoms in Fig. 4.4. For this atom, the s, p and d orbitals with the same value of n have very different energies.

The energy of each shell of electrons with the same principal quantum number always lies in the order s < p < d < f ..., but the subshells of a lower principal quantum number can overlap those of a higher number. The separation of the levels with the same value of n depends on the atomic number, Z. At large values of Z, there is no overlap of the sublevels, but, at intermediate Z, the order is found to be

$$1s < 2s < 2p < 3s < 3p < 4s < 3d < 4p < 5s < 4d < 5p.$$

It is this order which determines the behaviour of most atoms.

Atoms with different values of m_l only have different energies when a magnetic field is applied, as this enables a distinction to be made between orbitals oriented in different directions.

4.5 Pauli exclusion principle and the Aufbau principle

Apart from satisfying the Schrödinger equation, the wave functions of electrons have another condition that they must satisfy. This arises because the electrons are fermions, which have half-integral spins. The behaviour of fermions and bosons, such as photons, is very different. In general terms, bosons like to associate with each other, whereas fermions refuse to be crowded together. In 1924, Wolfgang Pauli expressed this property of fermions in more rigorous terms. He recognised that identical fermions cannot exist and, as a consequence, in a single atom, no two electrons can have all the same quantum numbers. This condition is called the *Pauli exclusion principle*. Its importance for chemists is that it requires electrons in atoms of different species to be arranged in quite different ways, giving rise to the diversity of chemical properties which the elements exhibit. Another consequence of this principle is that all acceptable wave functions for atoms and molecules must be anti-symmetric with respect to every pair of electrons, so that the wave functions change sign if the electrons are interchanged. This is an alternative formulation of the Pauli exclusion principle.

Having established the energy levels available to electrons in atoms, we need to find out how the electrons will fill the available levels. The rules that guide this filling-up process (in German, the *Aufbau prinzip*) are based on the Pauli exclusion principle. Each atomic orbital with a particular value of n, l and m_l can take two electrons, one with spin orientation $+1/2$ and one with spin orientation $-1/2$. We can represent each orbital by a circle and the electrons by ↑ or ↓, the direction of the arrow representing the + or − spin direction. The rules are:

(i) The orbitals are filled in order of increasing energy so that *those of lowest energy are filled first*. Each s orbital can take two electrons, each of the three p orbitals a pair and so on. The order we have established is given by the rule

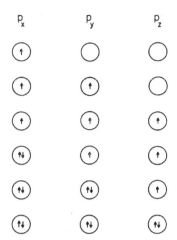

Fig. 4.5. The configurations that arise as electrons are added successively to fill three p orbitals.

that the orbital with the lowest value of $(n + l)$ has the lowest energy. If two orbitals have the same value of $(n + l)$, then the one with the lower n has the lower energy.

(ii) When electrons are added to orbitals of the same energy, such as a set of three p orbitals, *one electron will go into each of the available orbitals before any of the orbitals is filled by the addition of a second electron* (Fig. 4.5). This rule follows from the advantage of keeping the electrons apart; by going into different orbitals, the electron–electron repulsion is reduced and the arrangement is of lower energy than if the two electrons were in the same orbital. The half-filled subshell with three p electrons is a relatively stable arrangement. When an extra electron is added, the new configuration is less stable. The same general considerations control the way in which d orbitals are filled. Again, a relatively stable configuration is attained with five d electrons which form a half-filled subshell.

(iii) The relative orientation of the spins of the electrons is given by Hund's rule which states that the *electrons in different p orbitals (or other equivalent orbitals) will have their spins in the same direction* (until the orbitals begin to be filled with electrons of opposite spin), as shown in Fig. 4.5. This rule tells us that electrons in orbitals prefer to have their spins parallel but, by adopting opposing spins, they are able to pack in lower energy levels without contravening the Pauli exclusion principle.

Following these rules, we can now perform the filling-up process starting with the elements with the lowest atomic numbers. The results are illustrated in Fig. 4.6. For example, we can write the configuration for the carbon atom, C, as $1s^2 2s^2 2p^2$, where

	1s	2s	2p$_x$	2p$_y$	2p$_z$	3s	Ionization Energy/ kJ mol^{-1}
H	⊕	○	○	○	○	○	1310
He	⊕	○	○	○	○	○	2322
Li	⊕	⊕	○	○	○	○	519
Be	⊕	⊕	○	○	○	○	900
B	⊕	⊕	⊕	○	○	○	299
C	⊕	⊕	⊕	⊕	○	○	1088
N	⊕	⊕	⊕	⊕	⊕	○	1408
O	⊕	⊕	⊕	⊕	⊕	○	1314
F	⊕	⊕	⊕	⊕	⊕	○	1682
Ne	⊕	⊕	⊕	⊕	⊕	○	2080
Na	⊕	⊕	⊕	⊕	⊕	⊕	498

Fig. 4.6. The electronic configuration of selected elements.

the superscript means that two electrons are in the same subshell (it is *not* a power). The chemically stable noble gases have completely filled shells; for example, helium with $1s^2$ and neon with $1s^2 2s^2 2p^6$. Secondly, elements with similar outer electrons, such as lithium with one 2s electron and sodium with one 3s electron, have very similar chemical properties. The fact that the outer electrons largely determine the properties of the elements enables us to understand the origins of the periodic table. The first row (or period, as the rows are often called) of the table has two elements, hydrogen and helium. This corresponds to the filling of the $n = 1$ shell with two 1s electrons. The next row consists, as we would expect, of eight elements arising from the filling of the 2s and 2p$_x$, 2p$_y$ and 2p$_z$ orbitals. We might expect 18 elements in the next row, filling the 3s, 2p$_x$, 2p$_y$, 2p$_z$ and the five d orbitals. However, in the periodic table, only eight elements appear in this row. This tells us that the 3d orbitals are not filled until after the 4s. In the fourth and fifth rows, apart from the elements that represent the filling of the 4s, 4p, 5s and 5p orbitals, we have the 3d and 4d orbitals filled. In the sixth and seventh rows, apart from the s, p and d subshells, we have the filling of the 4f and 5f subshells. Thus, the structure of the periodic table (Fig. 4.7) confirms that the order of the energy levels of the lighter elements is best represented by:

$$1s < 2s < 2p < 3s < 3p < 4s < 3d < 4p < 5s < 4d < 5p < 6s < 4f < 5d$$
$$< 6p < 7s.$$

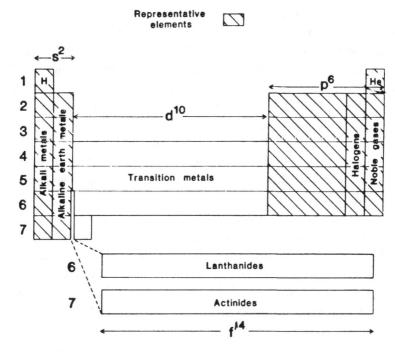

Fig. 4.7. The structure of the periodic table.

This pattern of levels is not always constant, but represents a sufficiently accurate basis for the interpretation of the periodic table. The fact that the 3d levels lie below the 4s levels enables us to understand the properties of the transition metals. As the outer shell remains constant, the elements show similar properties and, as electrons can be lost from both the 3d and 4s shells, the transition metals show variable valencies.

4.6 The shielding of outer electrons and atomic properties

The chemical properties of elements are determined, to a large degree, by the outer electrons, particularly the outermost s and p electrons. We call these electrons the *valence electrons*. Other electrons, such as those with lower principal quantum numbers and the d and f electrons, are referred to as the *core electrons*. The behaviour of the valence electrons is largely determined by how well the core electrons shield them from the attraction of the nucleus. In elements in the first few rows of the periodic table, the outer electrons are shielded almost completely by those electrons with smaller principal quantum numbers. Thus, the nuclear charge 'seen' by the first electron of a new shell is approximately the atomic number (the total positive charge on the nucleus) less approximately 1 unit for every *inner* electron. Thus, for sodium (configuration $1s^2 2s^2 2p^6 3s^1$) the 'effective' nuclear charge experienced by the outer electron is (very approximately) $11 - 10 = 1$, rather than the full nuclear charge, 11. Each electron added to the *same shell* provides only a little extra shielding (approximately one third). So, as we go across rows in the periodic table, the outer electrons of the elements experience a larger effective nuclear charge and are more tightly held. As we start a new row, the effective nuclear charge goes back to a little more than unity but the new electron is now in another shell and is further away from the nucleus. Thus, the electrons are less tightly held. The trend through the periodic table can be represented schematically as in Fig. 4.8. We find that many properties of the elements follow this general trend. Superimposed on this general pattern are features due to the filling and half filling of the subshells. The filling of the 2s orbitals in beryllium results in electrons which are more tightly-bound and the half-filled p subshell of nitrogen is also a relatively stable configuration.

We will return to discuss the topic of shielding in a more quantitative way later, but will first examine a number of properties that relate to the strength of binding of the outer electrons. A number of properties increase in magnitude with the strength of binding:

(i) *Ionization energies*. The energy required to remove one electron from an element in the gas phase (to an infinite distance away),

$$M(g) \rightarrow M^+(g) + e^-(g),$$

is known as the first ionization energy, I.

The values of I observed for a number of elements are given in Table 4.2 and are illustrated in Fig. 4.9. We see that, for the early rows (or periods) of the periodic table, the ionization energy increases through a row with two

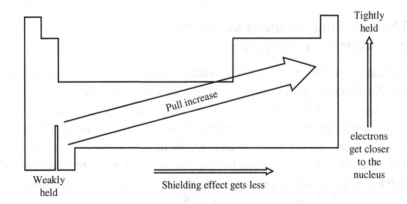

Fig. 4.8. The pull of the nucleus on the outer (valence) electrons.

Table 4.2. Properties of selected elements.

Element	Atomic number	Ionization energy kJ mol⁻¹	Electron affinity kJ mol⁻¹	Electronegativity	Atomic radius nm	Orbital Radius nm	Effective nuclear Charge (Slater)
H	1	1311	73	2.1	0.12	0.0529	1
He	2	2372			0.118	0.0291	1.7
Li	3	520.1	60	1	0.182	0.1586	1.3
Be	4	899.2		1.5		0.104	1.95
B	5	800.5	27	2	0.098	0.0776	2.6
C	6	1086.1	122	2.5	0.091	0.062	3.25
N	7	1402		3	0.092	0.0521	3.9
O	8	1314	141	3.5	0.074	0.045	4.55
F	9	1680.6	328	4	0.135	0.0396	5.2
Ne	10	2080			0.16	0.0354	5.85
Na	11	495.7	53	0.9	0.095	0.1713	2.2
Mg	12	737.5		1.2		0.1279	2.85
Al	13	577.4	44	1.5	0.16	0.1312	3.5
Si	14	786.3	134	1.8	0.132	0.1068	4.15
P	15		72	2.1	0.104	0.0919	4.8
S	16	999.3	200	2.5	0.127	0.081	5.45
Cl	17	1255.7	349	3	0.127	0.0725	6.1
Ar	18	1520			0.192	0.0659	6.75
K	19	418.6	48	0.8	0.133	0.2162	2.2
Br	35	1142.7	325	2.8	0.165	0.0851	7.6
Kr	36	1351			0.197	0.0795	8.25

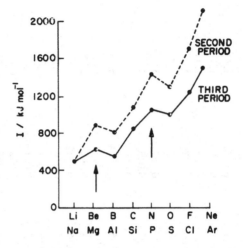

Fig. 4.9. Ionization energies.

discontinuities — one at the filling of the s subshell and one when the p subshell is half filled. Both represent configurations of extra stability.

(ii) *Electron affinity*. This is the energy *released* when a neutral atom combines with an electron to form an ion in the gas phase:

$$M(g) + e^- \rightarrow M^-(g).$$

The electron affinity, A, is a measure of the tendency of elements to form negative ions. They are hard to measure and values are not available for all elements. The major trend is such that elements with high electron affinities lie to the top right of the periodic table, with halogens having higher values than other elements. The rare gases have negative electron affinities, as an extra electron would have to go into a new shell.

(iii) *Electronegativity*. Electronegativity, χ, tells us when elements are likely to interact as ions and is a measure of the power of an atom to attract electrons to itself when forming a chemical bond. When elements of very different electronegativities combine they tend to form ionic bonds.

Consider the process

$$Na + Cl \rightarrow Na^+ + Cl^-,$$

which can be written as a sum of

$$Na \rightarrow Na^+ + e^-,$$

for which the energy change is I_{Na}, and

$$Cl + e^- \rightarrow Cl^-,$$

for which the energy change is A_{Cl}.

The net energy requirement for the overall reaction is $[I_{Na} - A_{Cl}]$. For the alternative process

$$Na + Cl \rightarrow Na^- + Cl^+,$$

the energy requirement is $[I_{Cl} - A_{Na}]$. The former reaction will be favoured if

$$I_{Na} - A_{Cl} < I_{Cl} - A_{Na}$$

or, rearranging, if

$$I_{Na} + A_{Na} < I_{Cl} + A_{Cl}.$$

Therefore, we define the electronegativity of an element as $\chi = (I + A)$. A greater electronegativity corresponds to a greater tendency to form negative ions. Thus, since $\chi_{Cl} > \chi_{Na}$, chlorine will exhibit the greater tendency to form negative ions.

Since electron affinities are difficult to measure, Linus Pauling, noting that the energy of bonds between two atoms was greater between atoms of very different electronegativities, employed this observation to develop a widely-used empirical scale of electronegativity.

The electronegativity of elements is closely related to their position in the periodic table. The elements at the top right of the table, where the outer electrons are most strongly bound, have the highest values of electronegativity and those elements at the bottom left of the table have the lowest values.

Other properties *decrease* in magnitude with the strength of the binding and are favoured in elements in which the outer electrons are highly screened:

(i) *Metallic character*: this includes the ability to conduct heat and electricity, the capacity to be hammered into shape and drawn into wires and is closely related to how loosely the outer electrons are bound. The properties of metals arise from the fact that electrons can escape from individual atoms and move freely through the solid. Non-metallic character is found at the top right corner of the periodic table. Metallic character decreases along the rows but increases down the groups. Between metals and non-metals lie the semiconductors, such as selenium and germanium, and, to a lesser extent, silicon; elements which are the basis of the modern electronics industry.

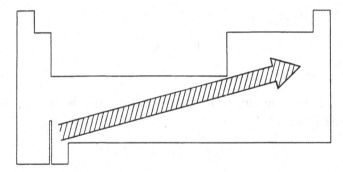

Fig. 4.10. Summary of trends through the periodic table: as strength of the binding of outer electrons increases along the direction of the arrow: 1) size *decreases*; 2) ionization energy (I) increases; 3) electron affinity (A) increases; 4) electronegativity (χ) increases; 5) positive ion formation *decreases*; 6) negative ion formation increases; 7) metallic character *decreases*.

(ii) *Atomic size*: this follows a similar trend. If the outer electrons are well-screened then they will be less tightly held and will be found further from the nucleus. Thus, the stronger the screening effect, the larger the atom. The variation in atomic size across rows is more regular than that of many other properties.

Thus, many general properties of the elements (see Table 4.2) are closely related to their position in the periodic table. The trends for a number of properties are illustrated in Fig. 4.10. These properties determine the reactivity, valencies and the types of chemical compound that the elements form. We can relate the detailed chemistry of these elements and their compounds to their electronic structure.

4.7 Estimating atomic properties

Many attempts have been made to devise methods which can be used to calculate atomic properties with reasonable accuracy. John Slater, in 1930, proposed a set of rules to estimate the effective nuclear charge experienced by electrons in different orbitals. The rules are set out in Table 4.3 and the values of the effective nuclear charge experienced by outer electrons of a number of elements are given in Table 4.3. Slater also proposed simpler functional forms for the wave functions of hydrogen-like atoms which have been widely used as the starting point for quantum-mechanical calculations of atomic and molecular properties. In addition, the effective nuclear charges established by Slater can be used in conjunction with the results we obtained in Section 3.8 to calculate the energy levels and orbital sizes of hydrogen-like atoms. They can also be used to estimate the polarisability, magnetic behaviour and other properties of the atoms of the elements.

Example: *Calculation of the ionization energy and the orbital radius of carbon.*

Table 4.3. Slater rules.

Slater assumed that the wave function of each electron in an atom could be expressed in terms of an effective quantum number, n^*, and an effective nuclear charge, $(Z - \sigma)$, where σ is a screening constant that represents the degree to which electrons are shielded from the true nuclear charge, Z.

To determine the screening constant, the electrons are divided into *groups*:

(1s) (2s, 2p) (3s, 3p) (3d) (4s, 4p) (4d) (4f) (5s, 5p) (5d).... *etc.*

The screening constant, σ, is made up of the following contributions:

  (i) Nothing from any *group* to the right of the one under consideration
 (ii) For s or p electrons:

    (a) a contribution of 0.35 from any other electron in the same *group* (except that for 1s electrons, the contribution is 0.30).
    (b) a contribution of 0.85 from each electron in the *group* immediately inside (that is, immediately to the left in the above list).
    (c) a contribution of 1.00 from all electrons still closer in.

(iii) For d or f electrons:

    (a) electrons in the same *group* contribute 0.35.
    (b) electrons in *groups* to the left, in the list above, of the one under consideration contribute 1.00.

The effective nuclear charges for the valence electrons of the elements in the first three rows of the periodic table, calculated using the Slater rules, are given in Table 4.2.

The effective principal quantum number, n^*, is related to the true principal quantum number, n, in the following manner.

n	1	2	3	4	5	6
n^*	1	2	3	3.7	4.0	4.2

Using these parameters Slater proposed modified forms of the hydrogen-like wave function in which the radial contribution was simplified. However, the modified parameters devised by Slater can be substituted in the equations derived from the solution of a hydrogen-like atom to estimate atomic properties. (Appendix 2)

Carbon has an electronic configuration $1s^2 2s^2 2p^2$. For the valence electrons, the screening of the nuclear charge provided by the inner electrons, σ, is given by (see Table 4.3)

$$Z = 6, \quad \text{and} \quad \sigma = 3 \times 0.35 + 2 \times 0.85 = 2.75,$$

giving

$$(Z - \sigma) = 3.25.$$

For the ion C^+, the configuration is

$$1s^2 2s^2 2p^1, \quad \sigma = 2 \times 0.35 + 2 \times 0.85 = 2.40, \quad \text{and} \quad (Z - \sigma) = 3.60.$$

The effective nuclear charge experienced by the inner 1s electrons is

$$(Z - \sigma) = 6 - 0.30 = 5.7.$$

The ionization energy of a valence electron of carbon is given by

$$I = E_{C^+} - E_C,$$

where

$$E_C = 2\,E(1s) + 4\,E(2s, 2p) \quad \text{and} \quad E_{C^+} = 2\,E(1s) + 3\,E(2s, 2p).$$

For each electron (see Section 3.8),

$$E = -\left[\frac{(Z - \sigma)}{n}\right]^2 \times 1312\,\text{kJ mol}^{-1},$$

so

$$E_C = -\left[2 \times (5.70)^2 + 4 \times \left(\frac{3.25}{2}\right)^2\right] \times 1312\,\text{kJ mol}^{-1} = -99112\,\text{kJ mol}^{-1}$$

and

$$E_{C^+} = -[2 \times (5.70)^2 + 3 \times (3.60/2)^2] \times 1312\,\text{kJ mol}^{-1} = -98006\,\text{kJ mol}^{-1},$$

giving

$$I = 1106\,\text{kJ mol}^{-1}.$$

The experimentally-determined value is $1086.1\,\text{kJ mol}^{-1}$.

The radius of electronic orbitals can be defined by the distance at which the electron density is a maximum. This is defined for hydrogen-like atoms as

$$R = \frac{n^2 a_0}{(Z - \sigma)},$$

where a_0 is the Bohr radius, $52.9\,\text{pm}$. For the valence electrons of carbon, we obtain

$$R = \frac{4a_0}{3.25} = 65\,\text{pm}.$$

The experimental value is $62\,\text{pm}$.

Calculations of this type can provide approximate values for a wide range of atomic properties. In addition, they have been the starting point for many atomic and molecular quantum-mechanical calculations. However, with the advent of modern electronic computers, calculations based on Slater orbitals are no longer the favoured method for determining the properties of atoms and, apart from the insight that the calculations provide, they are now only of limited practical value.

4.8   Solving the Schrödinger equation for atoms

The Schrödinger equation for electrons orbiting around a positively-charged nucleus can only be solved exactly when one electron is involved, as is the case for hydrogen atoms. When more than one electron is present, the exact solution of the equation is not possible. Even helium, with only two electrons, requires approximate methods, even though, in this case, the properties can be calculated to very high accuracy. For more complicated atoms, the factors determining the electrostatic potential energy are usually easy to formulate, so that the exact expression for the potential energy (the exact Hamiltonian, in the formalism of quantum mechanics) is usually known. Unfortunately, even with this knowledge, it is not possible to solve the Schrödinger equation exactly to obtain a wave function, ψ, and the allowed energies, E. It is usual to attempt to solve the Schrödinger equation by selecting approximate trial wave functions. Fortunately, for the ground states of atoms and molecules, a way exists to select the best of these functions. Every approximation will give a value for the energy of the lowest energy level which is *greater* than the true value. This is not surprising, as the true distribution of the electrons given by the correct wave function will be such that the energy is minimised. Any approximation will be inferior in this respect. This principle is called the *variational theorem* and has proved the key to many practical applications of the Schrödinger equation. It is common for trial wave functions to be expressed using a series of parameters which can be systematically varied until they provide the lowest value of the energy, given by

$$E = \int \psi^* \mathbf{H} \psi \, d\tau \left/ \int \int \psi^* \psi \, d\tau. \right.$$

The wave function and energy corresponding to these parameters is then the best approximation, though its accuracy is limited by the choice of trial wave functions.

An alternative method of solving the Schrödinger equation approximately is *perturbation theory*. This is employed when we are unable to solve the Schrödinger equation for the system under consideration, but can solve it exactly for a closely-related, simpler system. The Hamiltonian operator can be regarded as being composed of two parts, one for which we can solve the Schrödinger equation exactly, $\mathbf{H}^0$, and the extra perturbation term, $\mathbf{H}^1$, which reflects the system of interest:

$$\mathbf{H} = \mathbf{H}^0 + \mathbf{H}^1.$$

We can now write the energy of the system as

$$E = E^0 + E^1,$$

where E^0 is the energy of the original system and E^1 is the additional contribution due to the change or perturbation. In the same way, we can write the wave function as

$$\psi = \psi^0 + \psi^1.$$

Perturbation theory assumes that the major term in estimating the energy difference between the original system and the perturbed system can be written

$$E^1 = \int \psi^0 \mathbf{H}^1 \psi^0 \, d\tau.$$

Thus, the correct Hamiltonian operating on the approximate wave function allows us to make an improved estimate of the energy. Better approximations can be obtained by second-order perturbation theory which uses the improved wave function, ψ^1, to obtain yet more accurate values of the energy. However, the simplicity of first-order perturbation theory is lost.

We re-emphasise the fundamental principle which underpins perturbation theory by observing that, to a first approximation, the energy of the perturbed system is given by

$$E = \int \psi^0 \mathbf{H} \psi^0 \, d\tau = \int \psi^0 \mathbf{H}^0 \psi^0 \, d\tau + \int \psi^0 \mathbf{H}^1 \psi^0 \, d\tau = E^0 + E^1.$$

This represents the energy obtained by using the exact Hamiltonian, which we know, by allowing it to operate on the wave function corresponding to the unperturbed system. The physical implication of this equation is that the wave function of a slightly-perturbed system changes only little and can provide a sufficiently accurate approximate wave function that, used in conjunction with the correct Hamiltonian, can give a good estimate of the energy. The square of the wave function represents the distribution of the electrons and we could express the result more generally by stating that the energy of the system is more sensitive to changes in the potential energy function than is the configuration. It is useful to express it this way because we will find later that this conclusion provides a key to the solution of many non-quantum-mechanical problems in physical chemistry, not least the structure of liquids and the behaviour of mixtures.

4.9 The ground state of the helium atom

The approximate methods of solving the Schrödinger equation can be illustrated by their application to the helium atom. An exact solution is not possible, although, in this case, excellent approximations for the energy levels are available that are very close to the experimental values. The source of difficulty in this, as in all atomic and molecular problems, is that, although there is no problem in handling the interactions between each electron and the nucleus, the interaction between the electrons, the so-called r_{12} term, is hard to evaluate.

Experimental determinations for the helium atom give $-7623\,\text{kJ mol}^{-1}$ as the energy of the ground state and $2373\,\text{kJ mol}^{-1}$ as the first ionization energy. This is the difference between the energy of the helium ion He^+ and the energy of the ground state of the neutral helium atom. In Table 4.4, the results of the calculations which follow are compared with these experimental values. If we ignore the inter-electron repulsion, the energy of the helium atom is the same as that of two hydrogen-like atoms in which the electrons are exposed to a nuclear charge of 2. Thus, $E = 2\,Z^2/n^2 \times 1312\,\text{kJ mol}^{-1}$ and, since $Z = 2$ and $n = 1$, we obtain $10502\,\text{kJ mol}^{-1}$ for the ground state energy and $5250\,\text{kJ mol}^{-1}$ for the ionization energy. These estimates are so inaccurate as to be of little value when we consider that the typical energies we deal with in chemical reactions are on the order of $100\,\text{kJ mol}^{-1}$. This fact illustrates the difficulty chemists have when performing quantum-mechanical calculations. The properties in which we are interested are often calculated from the difference between two much larger quantities. A distinguished quantum chemist suggested that it was like measuring the weight of the captain of a ship by weighing the ship with him aboard and with him onshore!

The simplest way in which we can make allowance for the electron–electron interactions is by using Slater rules (see Table 4.3). For helium 1s electrons, the effective nuclear charge $(Z - \sigma) = 2.0 - 0.30 = 1.70$. Therefore,

$$E_{\text{He}} = -2 \times [(Z - \sigma)/n]^2 \times 1312\,\text{kJ mol}^{-1} = -2(1.70)^2 \times 1312\,\text{kJ mol}^{-1}$$

$$= -7583\,\text{kJ mol}^{-1}$$

and

$$E_{\text{He}^+} = -1 \times 2^2 \times 1312\,\text{kJ mol}^{-1} = -5248\,\text{kJ mol}^{-1}.$$

The ionization energy $I = E_{\text{He}+} - E_{\text{He}} = 2335\,\text{kJ mol}^{-1}$.

These values compare favourably with the experimental values of $7623\,\text{kJ mol}^{-1}$ and $2372\,\text{kJ mol}^{-1}$ (often quoted as $24.59\,\text{eV}$). This relatively good agreement is not unexpected, as the Slater rules were devised to give an effective nuclear charge in keeping with the experimental data.

If we apply the variational method using hydrogen-like wave functions, but allowing the effective nuclear charge to vary, we find that the energy becomes minimum when the nuclear charge is set at approximately 1.69 (very close to the value given by the Slater rules) and has the value $7477\,\text{kJ mol}^{-1}$ (Table 4.4). The corresponding ionization energy is $2226\,\text{kJ mol}^{-1}$, still almost $150\,\text{kJ mol}^{-1}$ in error.

Table 4.4. The ground state energy of the helium atom.

Method	Energy/ kJ mol^{-1}	Ionization energy/ kJ mol^{-1}
Neglecting electron-electron repulsion $E = 2 \times 4 \times E$ (hydrogen atom)	10502	5251
Slater screening calculation $(Z - \sigma) = 1.70$	7583	2335
Variational Calculations:		
Minimising energy by varying $(Z - \sigma)$. $(Z - \sigma) = 1.6875$	7477	2335
by varying both n and $(Z - \sigma)$. $n = 0.995$ and $(Z - \sigma) = 1.61162$.	7494	2242
Hartree-Fock limit	7513	2262
Hylleras (1929) 19 parameters	7623.5	2272.4
Perkeris (1959) 1078 parameters	7623.7	2372.7
Perturbation calculations:		
First-order (with electron-electron repulsion as the perturbation)	7220	1969
Second order	7634	2383
Thirteenth order	7623.7	2372.7
Observed value:	7622.6	2371.6

For references see *Physical Chemistry: A Molecular Approach*, D.A. McQuarrie and J.D. Simon, University Science Books, Sausalito, California, 1997, p. 278.

The use of more flexible wave functions, with many parameters that can be varied, enables extremely accurate values to be obtained. If we consider the inter-electron repulsion term as a perturbation, we find that the corresponding values are much improved, but the first order correction is still in error by some 400 kJ mol^{-1}. Second-order perturbation theory leads to an error of only 10 kJ mol^{-1}, some 0.4% in error. The essentially exact value has been obtained by applying perturbation theory up to the 13th order!

An alternative approach is to express the wave function of the helium atom in terms of the product of two highly-flexible, one-electron wave functions:

$$\psi(r_1, r_2) = \psi(r_1)\psi(r_2).$$

The procedure is to guess the form of one wave function and calculate the average electric field and the energy that results from it. The second electron is then allowed to interact with this field and a new wave function generated. This process is repeated back and forth until the wave functions become self-consistent and the energy converges to a limit. It might be thought that, by using very general wave functions, this approach might converge on the exact answer. It does not, however, because each electron only interacts with the *averaged* position of the other electron and not its instantaneous position. The limit we reach by this procedure is called the *Hartree–Fock limit*. It gives a value for the helium ground state of 7513 kJ mol^{-1}.

The discrepancy between the true energy and that given by the Hartree–Fock limit is called the *correlation energy* in this case amounting to $-110\,\text{kJ}\,\text{mol}^{-1}$. Though the Hartree–Fock energy is some 99% of the correct, experimentally determined energy, the error is still too large for the calculation to be helpful to chemists. It should be pointed out for the sake of completeness that small extra contributions to the energy also arise from relativistic effects. However, these are generally very small and not of chemical significance.

If sophisticated wave functions, not broken down into single-electron wave functions, but containing the r_{12} term explicitly, are employed, then, using powerful computational techniques, essentially exact results can be obtained. One such calculation used wave functions for the helium atom defined by 1078 parameters! Helium is a particularly simple case and, when one proceeds to large atoms and molecules, the scale of the computational problems make it more difficult to obtain sufficiently accurate results. However, such is the power of modern computers that standard programs are available which enable chemists to compute quite accurate atomic and molecular properties.

4.10   Summary of key principles

The solution of the *Schrödinger equation* for many-electron atoms leads to wave functions characterised by three *quantum numbers*: n, the principal quantum number, which plays the major part in determining the energy, l, the angular momentum quantum number, which defines the shape of the atomic orbital and m_l, the magnetic quantum number, which defines their orientation in space. An additional quantum number, m_s, arises from *electron spin*. Electron spins can have two orientations represented by $m_s = \pm 1/2$. When more than one electron is present, the energy levels depend on n and l (and, in the presence of a magnetic field, also on m_l).

The manner in which the atomic energy levels are filled is determined by the *Pauli exclusion principle*, which states that, in any atom, no two electrons may have all the same quantum numbers. If orbitals of the same energy are available, electrons will occupy different orbitals and each electron will have the same spin orientation. However, if orbitals of significantly different energy are available, then the two electrons will occupy the orbital of lowest energy and, following the Pauli principle, will have opposite spins.

The properties of elements, and the way in which they are related to their position in the periodic table, are largely determined by the manner in which the outer electrons are bound by the positively charged nucleus. This depends both on the strength of the nuclear charge and the degree to which it is shielded by the inner electrons.

It is generally not possible to solve the Schrödinger equation exactly for systems of interest to chemists, but approximate methods are available for its solution. The *variational method* calculates the energy arising from various approximate wave functions in the knowledge that the wave function which leads to the lowest energy will be the most accurate. The *perturbation method* is employed when it is possible to solve the Schrödinger equation for a simple system closely related to the one for which a solution is required. Then, the wave function for this simple system can be used in conjunction with the exact Hamiltonian to provide an estimate of the energy of the system under investigation.

Problems

(Use the Slater rules, given in Table 4.3, for the problems below.)

(1) Estimate the effective nuclear charge experienced by a valence electron in the magnesium atom.
(2) Estimate the effective nuclear charge and calculate the ionization potential of the lithium atom.
(3) Estimate the distance from the nucleus at which the valence electron density is a maximum for the sodium atom.

5

Chemical Bonding and Molecular Structure

Having observed the difficulty of solving the Schrödinger equation for atoms with more than one electron, it will come as no surprise that its application to molecules presents formidable difficulties and there is no question of exact solutions. However, through the use of approximate methods, the application of the Schrödinger equation has enabled considerable insight to be gained into the nature of chemical bonding and has provided a framework with which to understand the behaviour of electrons in molecules.

5.1 The chemical bond — a historical digression

It is instructive to consider early approaches to chemical bonding which have provided important concepts that are still of value today. The first coherent picture of chemical bonding was developed by G. N. Lewis (and others) in the years following 1916 in terms of a simple rule:

'Atoms tend to gain, lose or share electrons in order to acquire completely filled shells like the inert gases.'

In fact, apart from hydrogen, for which two electrons suffice, the rule suggested that, for the elements of the first two rows of the periodic table, each atom within a stable molecule needs to acquire eight valence electrons. If two hydrogen atoms are in close proximity, they can *share* electrons to form a covalent bond in which each has reached a helium structure, $1s^2$. An alternative process can occur by which, for instance, sodium can *lose* an electron to become Na^+ ($1s^22s^22p^6$) and chlorine can *acquire* an electron to become Cl^- ($1s^22s^22p^63s^23p^6$). Both atoms attain inert gas structures (but with opposite electric charge) and form an ionic bond, represented by Na^+ --- Cl^-.

Lewis proposed that covalent bonds were formed by pairs of shared electrons and could be illustrated by representing each electron as a dot so that a hydrogen molecule becomes H:H. Covalent bonds can be single, as in this example, but, when more than two electrons are shared, the bond can become a double or even a triple bond. Thus, oxygen atoms need to share four valence electrons to complete their outer shells and they combine to form molecular oxygen in which the atoms are joined by a double bond,

$$::O::O::$$

Nitrogen atoms need to share six electrons to attain an inert gas structure and thus require a triple bond to form the nitrogen molecule

$$:N:::N:$$

The Lewis theory works well for elements in the first two rows of the periodic table but breaks down where atoms have electrons in 3d orbitals. Even in the second row, a few stable compounds of boron, such as BF_3, where the boron atom shares only six electrons, fail to satisfy the eight electron rule.

Lewis diagrams can be used to predict the shapes of molecules if we assume that bonding electrons and any non-bonding lone pair electrons repel each other. Such repulsion will lead CH_4 to assume a tetrahedral configuration with bonds at an angle of 109.5° to each other. If we consider NH_3, with three bonds and one lone pair of electrons, the angle is distorted to 107.3°, and in water, where the oxygen has two lone pairs of electrons, the angle is 104.5°. From this, we can deduce that pairs of bonding electrons repel each other and that *lone pairs of electrons repel even more strongly than the bonding electrons*. This method of understanding molecular structure has the grand name of the *valence shell electron pair repulsion (VSEPR) model* and, considering its simplicity, is remarkably successful. One should note that, in this model, multiple bonds count only as a single bond for the purposes of determining molecular shape.

For some molecules, more than one Lewis structure can be written. A simple example of this is benzene, the skeleton of which Lewis represented as an alternating series of single and double carbon–carbon bonds. The structure can be written in two ways and Lewis assumed that there could be rapid interchange between the two structures, which could explain the equivalence of all the carbon–carbon bonds in benzene. The two structures were termed *resonance structures* and the resulting combination a *resonance hybrid*. It was noted that, where this process of resonance was thought to occur, the molecules tended to be more stable than would be expected on the basis of either of the individual resonance structures. Resonance proved a necessary, but clumsy, explanation for a phenomenon that could not be understood until the advent of quantum mechanics. The alternating patterns of single and double bonds that occur in the Lewis representation of aromatic molecules and some linear hydrocarbons are called *conjugated* systems.

Though the Lewis theory provided many useful insights into atomic reactivity and molecular structure, we are now in a position to develop a much more thorough understanding of the way electrons are distributed within molecules and to see how this provides an explanation of the structure and behaviour of molecules.

5.2 Valence bond theory

It is not surprising that the first attempt to provide a more rigorous quantum-mechanical explanation of valency closely followed features of the Lewis model. In 1927, immediately after Schrödinger introduced the wave equation, Heitler and London provided a quantum-mechanical model of covalent bonding for the hydrogen molecule. The electrostatic interactions within a hydrogen molecule are illustrated in Fig. 5.1. Each contributes a term of the form e^2/r to the total electrostatic potential energy. The Hamiltonian can be written

$$H = -\frac{h^2}{8\pi^2 m}(\nabla_1^2 + \nabla_2^2) - \frac{e^2}{4\pi\varepsilon_0}\left(\frac{1}{r_{A1}} + \frac{1}{r_{B1}} + \frac{1}{r_{A2}} + \frac{1}{r_{B2}} - \frac{1}{r_{12}} - \frac{1}{R}\right),$$

where A and B refer to the nuclei and 1 and 2 to the electrons. We can assume that electronic motion is so rapid when compared with molecular vibrations that we may regard the inter-nuclear distance R as constant. This is known as the *Born–Oppenheimer approximation* and is sufficiently accurate for most purposes of interest to chemists.

It is not possible to derive the exact wave function from first principles. When the two nuclei are far apart, the atoms can be accurately described by the two separate atomic wave functions. The Hamiltonian is then the sum of the separate atomic Hamiltonian functions and the total energy is simply that of two hydrogen atoms, $E_A + E_B$. However, as the nuclei come closer together, additional interactions occur, with the electrons on one atom interacting with the opposite nucleus, and the two electrons and the two nuclei repelling each other.

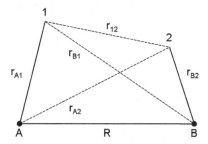

Fig. 5.1. The coordinates for the hydrogen molecule: A and B represent the nuclei of the hydrogen atoms and 1 and 2 the electrons.

The simplest approximate wave function that we can write for a hydrogen molecule is the product of the two 1s wave functions of atomic hydrogen:

$$\psi = \psi_A(1)\psi_B(2).$$

When the Hamiltonian given above operates on a wave function of this form, the resulting energy can be written

$$E = 2E_H + \left(\frac{e^2}{4\pi\varepsilon_0}\right)\left[\frac{1}{R} + \int\int\left(\frac{1}{r_{12}}\right)\psi_A^2(1)\psi_B^2(2)d\tau_1 d\tau_2\right.$$

$$\left. - \int\left(\frac{1}{r_{A2}}\right)\psi_B^2(2)d\tau_2 - \int\left(\frac{1}{r_{B1}}\right)\psi_A^2(1)d\tau_1\right]$$

where $\int\cdots d\tau$ implies integration over all the coordinates represented in the integral. When the two atoms are far apart, this wave function leads to the solution for independent hydrogen atoms, as given by the first term. The second term represents the energy arising from the repulsion of the positively charged nuclei. The third term arises from the repulsions between the electron associated with nucleus A and that associated with nucleus B. The fourth and fifth terms represent the interactions between the electrons on one atom with the nucleus of the other atom. It is necessary for these last two terms, with their negative contribution to the potential energy, to dominate the other terms of opposite sign if the two atoms are to bind to form a stable molecule. This occurs when the wave functions reflect a build up of electrons between the nuclei to provide the 'glue' that holds the nuclei together to form a stable molecule. The terms other than the first contribute to what is called the *Coulomb integral, J*. Evaluation of the integral shows that our approximate wave function does lead to a stable hydrogen molecule, but the binding energy is only some 5% of that observed experimentally (Fig. 5.2).

Heitler and London recognised that the weakness of this approach was that it failed to identify that *electrons are indistinguishable* and that each electron can be associated with either nucleus A or nucleus B. This means that the wave function $\psi_A(2)\psi_B(1)$ is as equally acceptable as $\psi_A(1)\psi_B(2)$. To incorporate the indistinguishability of the electrons, they proposed a wave function of the form

$$\psi = \psi_A(1)\psi_B(2) + \psi_A(2)\psi_B(1).$$

This leads to an additional contribution to the energy in the form of an integral

$$K = \left(\frac{e^2}{4\pi\varepsilon_0}\right)\left[\frac{S^2}{R} - S\int\frac{1}{r_{B1}}\psi_A(1)\psi_B(1)d\tau_1 - S\int\frac{1}{r_{A2}}\psi_A(2)\psi_B(2)d\tau_2\right.$$

$$\left. + \int\int\frac{1}{r_{12}}\psi_A(1)\psi_B(2)\psi_B(1)\psi_A(2)d\tau_1 d\tau_2\right],$$

where $S = \int\psi_A(1)\psi_B(1)d\tau_1$.

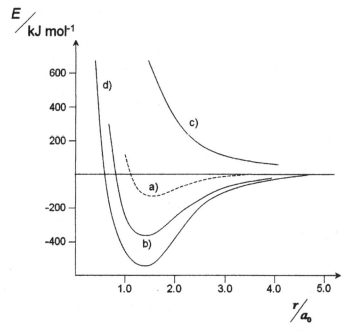

Fig. 5.2. The energy of a hydrogen molecule as a function of the separation of the atomic nuclei:
(a) the result obtained from a wave function that does not allow for the indistinguishably of electrons;
(b) Heitler–London result for the ground state; (c) Heitler–London result for the first excited (triplet)
state; (d) experimentally observed value.

K is called the *exchange integral* and S the *overlap integral*. The energy of the
hydrogen molecule expressed in terms of these integrals becomes

$$E = 2E_{\mathrm{H}} + \frac{(J + K)}{(1 + S^2)},$$

where *both J and K are negative quantities.*

This allowance for the indistinguishability of electrons produces a dramatic
improvement, with the incorporation of the new wave function leading to an estimate
of the bond energy of the hydrogen molecule of some two-thirds of the correct value
(see Table 5.1, Fig. 5.2). The overwhelming majority of this energy comes from the
exchange term. It is important to note that the 'exchange' of the electrons is not a
physical process, but simply the introduction of additional flexibility into the wave
function. (It was a common feature of the early years of quantum mechanics that
nonexistent dynamical physical processes were employed to explain the benefits
arising from improved wave functions.)

Even though it does not give an accurate value of the binding energy of the
hydrogen molecule, the valence bond treatment of Heitler and London provides a
good insight into the nature of chemical bonding. The wave function they employed

Table 5.1. Bond energy and bond length of the hydrogen molecule.

Method	Wavefunction	Bond energy/ kJ mol^{-1}	Bond length/ 10^{-10} m
Valence-bond: without exchange	$\psi_A(1)\psi_B(2)$	24.0	0.900
Valence-bond:	$\psi_A(1)\psi_B(2)+\psi_A(2)\psi_B(1)$	303	0.870
Valence-bond: with screening $Z_{eff} = 1.166$		363	0.740
Molecular orbital:	$[\psi_A(1) + \psi_B(1)][\psi_A(2) + \psi_B(2)]$	259	0.850
Molecular orbital: with screening $Z_{eff} = 1.24$		335	0.730
STO — 3G	See Section 5.11	307	0.710
Hartree–Fock limit		347*	0.730
James and Coolidge with r_{12} terms		453	0.740
Configuration interaction		457	0.742
Observed value		457	0.742

*The *correlation energy*, the difference between the Hartree–Fock limit and the observed value, is 110 kJ mol^{-1}.

is symmetrical in that the exchange of electrons leaves the wave function unchanged. The Pauli principle tells us that the overall wave function for electrons must be anti-symmetric. For this to be achieved, the contribution of the electron spins to the overall wave function must be anti-symmetric, requiring the spins to be of opposite sign. This is how the pairing of electrons of opposite spins as a requirement for bonding can be seen to arise. It results from the symmetry restrictions imposed by the Pauli principle. However, there is another wave function that is an equally acceptable representation of a hydrogen molecule, namely

$$\psi = \psi_A(1)\psi_B(2) - \psi_A(2)\psi_B(1).$$

In this case, the wave function is anti-symmetric with respect to the exchange of electrons and, for the overall wave function to satisfy the Pauli exclusion principle, the electron spin contribution must be symmetric, requiring that the electron spins are of the same sign and parallel to each other. The energy of this state is given by

$$E = 2E_H + \frac{(J - K)}{(1 - S^2)}$$

and the negative value of K makes a positive contribution to the energy, leading to an unstable excited state (Fig. 5.2). It is referred to as a *triplet state*, because the electrons, when parallel, can give rise to three energy levels of similar energy.

5.3   Molecular orbitals

The valence bond theory starts from the standpoint that the individual atoms retain their identity and character as the nuclei come closer together. An alternative way of looking at how binding occurs is called the *molecular orbital theory*. In this model, the electrons are thought of as belonging to the molecule as a whole, each occupying orbitals, defined by the wave function for a single electron, which are characteristic of the molecule rather than of the interacting atoms. These are *molecular orbitals*.

The wave functions of the molecular orbitals are most commonly constructed by a linear combination of the orbitals of the individual atoms. This is called the *linear combination of atomic orbitals (LCAO) approximation* and it provides the starting point for most molecular orbital calculations. For simplicity, we will consider the case of a molecule formed of only two atoms labelled 1 and 2, though the procedure is equally applicable to polyatomic molecules. The wave function of the molecular orbital is

$$\psi = c_1\psi_1 \pm c_2\psi_2.$$

Substituting in the equation

$$E = \frac{\int \psi \mathbf{H}\psi d\tau}{\int \psi^2 d\tau},$$

we obtain $E = \int (c_1\psi_1 + c_2\psi_2)\mathbf{H}(c_1\psi_1 + c_2\psi_2)d\tau / \int (c_1\psi_1 + c_2\psi_2)^2 d\tau$.
To simplify the equations we use the notation
$H_{ij} = \int \psi_i \mathbf{H}\psi_j d\tau$ and, for the overlap integrals, $S_{ij} = \int \psi_i\psi_j d\tau$.
E can then be written

$$E = \frac{(c_1^2 H_{11} + 2c_1 c_2 H_{12} + c_2^2 H_{22})}{(c_1^2 S_{11} + 2c_1 c_2 S_{12} + c_2^2 S_{22})}.$$

If we apply the variational principle to find the values of c_1 and c_2 which minimise the energy, we obtain a pair of linear equations known as *secular equations*,

$$c_1(H_{11} - E\,S_{11}) + c_2(H_{12} - E\,S_{12}) = 0$$

and

$$c_1(H_{12} - E\,S_{12}) + c_2(H_{22} - E\,S_{22}) = 0.$$

We wish to solve these equations to obtain E and the coefficients c_1 and c_2. They are solved using the Rayleigh–Ritz method, which predated quantum mechanics. For homonuclear diatomic molecules, such as hydrogen, c_1 and c_2 are both equal to $(1/2)^{1/2}$. The permitted values of E are the roots of what is called a *secular determinant* arising from these equations. (H_{ij} and S_{ij} are known as *matrix elements*.)

$$\begin{vmatrix} H_{11} - E\,S_{11} & H_{12} - E\,S_{12} \\ H_{12} - E\,S_{12} & H_{22} - E\,S_{22} \end{vmatrix} = 0$$

Expanding the determinant and solving the resulting quadratic equation gives two values of E. The method is easily extended to allow for a larger number of wave functions contributing to the linear combination and is the basis of many quantum-mechanical calculations of molecular properties. H_{ii} (which is $\int \psi_i \mathbf{H}_{ii} \psi_i d\tau$) are called (yet other) *Coulomb integrals*. H_{ij}($\int \psi_i \mathbf{H}_{ij} \psi_j d\tau$) are called *resonance integrals*, but the term resonance does not refer to any physical process. S_{12} is referred to as the *overlap integral*, as it is a measure of the extent to which the orbitals overlap. In evaluating the secular equations, many different levels of approximation are available, depending on the nature of the problem and the desired level of accuracy.

We can now apply the molecular orbital method to the simplest case, that of a hydrogen molecule. In order to evaluate the integrals arising in the secular determinant we make a number of further approximations. First, we can simplify the notation by writing the integrals

$H_{11} = H_{22} = \alpha$ $H_{12} = \beta$ and making the approximations

$S_{11} = S_{22} = 1$ and $S_{12} = 0$ giving

$$\begin{vmatrix} \alpha - E & \beta \\ \beta & \alpha - E \end{vmatrix} = 0.$$

Expanding the determinant leads to the equation $\alpha^2 - 2\alpha E + E^2 - \beta^2 = 0$, for which the solutions are

$$E = \alpha \pm \beta.$$

Though we have assumed that the overlap integral is zero, for light atoms this is a poor approximation, and it is sometimes necessary to take the value of the overlap integral into consideration, making the resulting expression in the energy a little more complicated,

$$E = \frac{(\alpha \pm \beta)}{(1 \pm S)}.$$

The wave function, $\psi = \psi_1 + \psi_2$, when both contributing wave functions have the same sign, corresponds to a significant electron concentration in the space between the nuclei and leads to the formation of a stable molecule. It is identified as a *bonding orbital* (Fig. 5.3). The second wave function, $\psi = \psi_1 - \psi_2$, when the wave functions

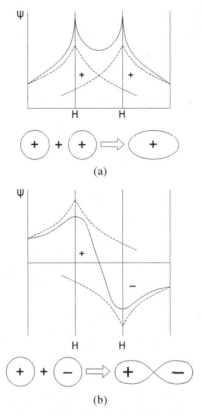

Fig. 5.3. Molecular orbitals in the hydrogen molecule. The broken lines indicate the 1s atomic wave functions of the hydrogen atoms. The solid lines indicate the molecular wave functions for the hydrogen molecule: (a) bonding molecular orbitals, $\psi = \psi_{1s}(A) + \psi_{1s}(B)$; (b) anti-bonding molecular orbitals, $\psi = \psi_{1s}(A) - \psi_{1s}(B)$.

have opposite signs, leads to a low concentration of electrons in the inter-nuclear space. This does not lead to an attraction between the atoms and the orbital is called an *anti-bonding orbital*. Electrons in the bonding orbital have an energy $\alpha + \beta$ and those in the anti-bonding orbital $\alpha - \beta$. Both α and β are negative quantities. To a reasonable approximation, α can be identified with the energy of a separate hydrogen atom. The additional energy contributed by an electron in the bonding orbital, that is, the contribution to the bonding energy from that electron, is

$$\Delta E = E - E_H = \beta.$$

In a similar way, the contribution to the energy of the molecule from an electron in the anti-bonding orbital is $\Delta E = E - E_H = -\beta$, and, as β is a negative quantity, no bond results.

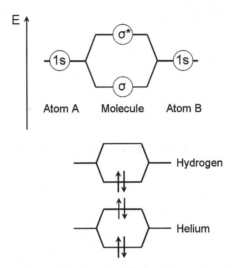

Fig. 5.4. The molecular orbitals for hydrogen (H_2) and helium (He_2). The asterisk indicates an anti-bonding orbital.

Having established the molecular orbitals, we can now feed in electrons in a similar way to the filling of atomic orbitals (Fig. 5.4). (The labelling of molecular orbitals is explained in the next section.) If we add a single electron to the two nuclei it will form the hydrogen molecular ion, H_2^+. The electron will go into the level with the lowest energy and, as this is a bonding orbital, the hydrogen molecular ion should be a stable species. This is found to be the case, and H_2^+ has been shown to exist spectroscopically and has a binding energy of approximately $269\,\mathrm{kJ\,mol^{-1}}$ at an inter-nuclear distance of 106 pm. If we feed in a second electron we would expect, in this simple model, twice the binding energy. In fact, this is not too far from being the case, as the binding energy of a hydrogen molecule is $455\,\mathrm{kJ\,mol^{-1}}$ at an inter-nuclear separation of 74 pm. Unfortunately, this measure of agreement comes about through a cancellation of errors arising from the fact that we have not taken account of the electron–electron repulsion that would occur when we added a second electron to the same molecular orbital.

We can apply similar reasoning to the case of a diatomic helium molecule with four electrons. Two will fill the bonding orbital and the remaining two the anti-bonding orbital. The result is no net bonding. The helium molecular ion, He_2^+, has three electrons, two of which will be in the bonding orbital and one in the anti-bonding orbital, suggesting it is a stable species. We can define a quantity called the *bond order*, B, defined in terms of the number of electron pairs in bonding orbitals less the number in anti-bonding orbitals:

$$B = \frac{1}{2}[n_e(\text{bonding}) - n_e(\text{anti-bonding})].$$

Table 5.2. Properties of diatomic molecules from the first two rows of the periodic table.

Molecule	Bond order	Bond energy/ kJ mol^{-1}	Bond length/ pm
H_2^+	$\frac{1}{2}$	269	106
H_2	1	457	74
$He_2^{!}$	$\frac{1}{2}$	241	108
He_2	0	0	—
Li_2	1	105	267
Be_2	0	0	—
B_2	1	290	159
C_2	2	600	124
N_2	3	942	110
O_2	2	490	121
F_2	1	154	140

The results for diatomic hydrogen and helium and their molecular ions, and for some other common elements, are given in Table 5.2.

We have studied the stability of the hydrogen molecule using two quite different models, the molecular orbital model and the valence bond model. Both show that a strong bond exists between the atoms but neither gives an accurate estimate of the bond energy. The reasons for this failure are different in each case. The molecular orbital model allows the two electrons to be spread uniformly in the space about the nuclei. It takes no account of the electron–electron repulsion, which tends to keep the electrons apart from each other. On the other hand, the valence bond model is based on a wave function that does not allow both electrons to be close to the same nucleus at the same time. It over-estimates the effect of electron repulsion. As would be expected from the variational principle, the predictions of both theories are higher than the true energy and underestimate the binding energy of the hydrogen molecule. Both theories can be improved by adding additional terms to the wave function.

5.4 Homonuclear diatomic molecules

Combining two atomic s orbitals creates a molecular orbital that is symmetrical about the inter-nuclear axis (Fig. 5.3). These orbitals, which have zero angular momentum about the inter-nuclear axis, are designated σ orbitals. The molecular orbital would also have the same symmetry if an s orbital on one atom combined

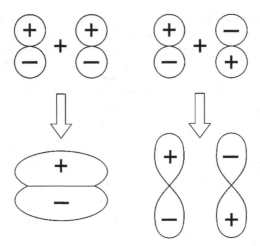

Fig. 5.5. The formation of π-bonds from atomic p_z orbitals: (a) bonding π orbitals (u); (b) anti-bonding π* orbitals (g).

with the lobe of a p orbital pointing along the inter-nuclear axis, usually identified as the p_z orbital. However, if the molecular orbital results from two p orbitals with lobes at right angles to the inter-nuclear axis, p_x and p_y, a molecular orbital of different symmetry arises with one unit of angular momentum about the inter-nuclear axis (Fig. 5.5). These orbitals are termed π orbitals and, as with σ orbitals, can be either bonding or anti-bonding. We can see a parallel between the σ orbitals of molecules and the s orbitals of atoms and between π orbitals and atomic p orbitals. Molecular orbitals are usually characterised both by their symmetry and by the atomic orbitals from which they may be thought to arise. The bonding and anti-bonding molecular orbitals of hydrogen, which we investigated earlier, would be designated σ1s and σ*1s. (The asterisk is used to denote anti-bonding orbitals.) Alternatively, one can note that the bonding σ orbital is symmetrical with respect to inversion through the centre of symmetry of the molecule (see Appendix 3). Such orbitals are labelled g, where g stands for *gerade*, the German word for *even*. The anti-bonding σ orbital is anti-symmetrical with respect to inversion through the centre of symmetry and is labelled *ungerade*, σ_u. Whereas bonding σ orbitals are symmetrical with respect to inversion, for π orbitals, the bonding orbitals are ungerade and the anti-bonding orbitals, gerade (Fig. 5.5). With more sophisticated molecular orbital calculations, the resulting orbitals cannot always be associated so clearly with atomic orbitals and it is common to use a simpler notation in which just the symmetry of the orbital, σ or π, and its number, counting from the lowest energy, are indicated.

The molecular orbitals arising from the combination of the atomic orbitals of the atoms in the first row of the periodic table are illustrated in Fig. 5.6. Because there

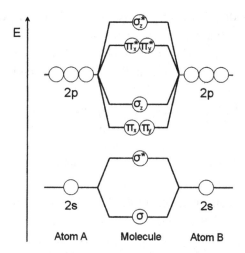

Fig. 5.6. Molecular orbitals for homonuclear diatomic molecules.

are two directions in which the p orbitals can lie at right angles to the inter-nuclear axis, π orbitals are degenerate. In a very similar way to atomic orbitals, we can list molecular orbitals in order of increasing energy and expect them to be filled in a similar way. Each orbital is capable of taking two electrons with opposite spin, as required to satisfy the Pauli exclusion principle. For the diatomic molecules formed from the lighter elements the order in which the orbitals are filled is generally

$$\sigma 1s, \ \sigma^* 1s, \ \sigma 2s, \ \sigma^* 2s, \ \pi 2p_x = \pi 2p_y, \ \sigma 2p_z, \ \pi^* 2p_x = \pi^* 2p_y, \ \sigma^* 2p_z.$$

The electronic configurations of valence electrons of the diatomic molecules formed by the elements of the second row of the periodic table are given in Table 5.3.

Table 5.3. Electronic Configurations of Diatomic Molecules: For the ground states of the elements of the second row of the periodic table. 1s orbitals of the atoms are not included. Note that for nitrogen, oxygen, flourine and neon the $(\sigma 2p_z)^2$ orbital has a lower energy and lies below the π orbitals.

$Li_2(\sigma 2s)^2$

$Be_2(\sigma 2s)^2(\sigma^* 2s)^2$

$B_2(\sigma 2s)^2(\sigma^* 2s)^2(\pi 2p_x)(\pi 2p_y)$

$C_2(\sigma 2s)^2(\sigma^* 2s)^2(\pi 2p_x)^2(\pi 2p_y)^2$

$N_2(\sigma 2s)^2(\sigma^* 2s)^2(\sigma 2p_z)^2(\pi 2p_x)^2(\pi 2p_y)^2$

$O_2(\sigma 2s)^2(\sigma^* 2s)^2(\sigma 2p_z)^2(\pi 2p_x)^2(\pi 2p_y)^2(\pi^* 2p_x)(\pi^* 2p_y)$

$F_2(\sigma 2s)^2(\sigma^* 2s)^2(\sigma 2p_z)^2(\pi 2p_x)^2(\pi 2p_y)^2(\pi^* 2p_x)^2(\pi^* 2p_y)^2$

$Ne_2(\sigma 2s)^2(\sigma^* 2s)^2(\sigma 2p_z)^2(\pi 2p_x)^2(\pi 2p_y)^2(\pi^* 2p_x)^2(\pi^* 2p_y)^2(\sigma^* 2p_z)^2$

Lithium has two bonding electrons in the σ orbital arising from the 2s atomic orbitals (Table 5.2). The bond in Li_2 is weaker and longer than the bond in molecular hydrogen, as 2s orbitals of lithium are more diffuse than the 1s orbitals of hydrogen. Be_2 has no net bonding electrons and does not exist as a stable species. B_2 has two bonding electrons in the two π orbitals. One electron goes into each of the orbitals and the spins are parallel. The molecular orbital theory predicts that B_2, because of its unpaired spins, would be paramagnetic, which is in agreement with experimental observation. Carbon and nitrogen have no unpaired electrons and the bond order of C_2 is two and that of N_2, three. We see that molecular oxygen also has two unpaired electrons and would also be expected to be paramagnetic. This is indeed the case and the prediction is a considerable success for molecular orbital theory, as it is not explained by either the valence bond or the Lewis electron-pair theories. Continuing through the row, F_2 has two net bonding electrons and Ne_2 has zero bond order and does not exist as a stable molecule.

5.5 Heteronuclear molecules

Molecular orbital theory can be extended to heteronuclear diatomic molecules, but now the contribution of the atomic orbitals will not be equal and, when we write the wave function as $\psi = c_1\psi_1 + c_2\psi_2$, c_1 and c_2 will no longer be equal. For atoms that are similar, as is the case, for example, with CO and NO, we can treat the diatomic molecules in the same way that we treated homonuclear molecules. In the case of nitric oxide, for example, oxygen contributes six valence electrons and nitrogen five, so that when we fill the molecular orbitals, one electron remains unpaired, telling us that NO is paramagnetic.

$$NO: \quad (\sigma\,1s)^2(\sigma^*\,1s)^2(\sigma\,2s)^2(\sigma^*2s)^2(\sigma\,2p_z)^2(\pi\,2p_x)^2(\pi\,2p_y)^2(\pi^*\,2p_x)$$

The bond order is $2\frac{1}{2}$, halfway between the oxygen and nitrogen values. However, when the atoms in the molecules differ significantly in their electronegativity (see Section 4.6), the electrons will tend to reside closer to one nucleus than the other. This asymmetry is said to confer 'ionic character' and we can consider the bond to be part covalent and part ionic. In hydrogen fluoride, the 1s orbitals of the hydrogen and the 2s orbitals of the fluorine are so different in energy that they cannot interact (Fig. 5.7). Only the 2p orbital on the fluorine is close enough in energy to be able to combine with the 1s orbital of hydrogen to form a bonding and an anti-bonding σ molecular orbital. However, the 2p orbitals of the fluorine lie some $370\,kJ\,mol^{-1}$ below the 1s hydrogen orbital. This means that there will be more fluorine 2p character in the bond and the charge distribution will not be symmetrical, with the bonding electrons lying closer to the fluorine nucleus than the hydrogen. The fluorine atom will have

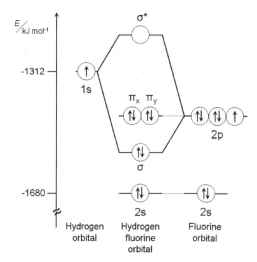

Fig. 5.7. The molecular orbitals of hydrogen fluoride. (The 1s orbital of fluorine is not shown.)

an excess of negative charge and the molecule will have a nonzero dipole moment. The $2p_x$ and $2p_y$ orbitals of the fluorine have zero overlap with the 1s electrons of the hydrogen and are termed *non-bonding* orbitals, as is the fluorine 2s orbital.

Molecular orbital theory can also be extended to polyatomic molecules, but now we might expect to construct orbitals from a combination of the orbitals of the valence electrons of all atoms in the molecule. Such orbitals can be employed when studying aromatic compounds, but, for most other types of molecule, it is simpler to construct localised two-centre molecular orbitals that can represent individual chemical bonds. For example, in describing water, we could regard the bonds between the hydrogen and the oxygen atoms as involving the 1s orbitals of the hydrogen and two of the 2p orbitals of the oxygen. The fact that the bond angle in water, 104.5°, is little more than 90° indicates that the geometry of the water molecule is, indeed, largely determined by the shape of the p orbitals. We assume, to a good approximation, that bonds are independent of each other and, if we replace one of the hydrogen atoms by another group, the remaining bond will be substantially unchanged. Thus, each bond has its own characteristic features, such as energy and length, which are determined by the wave function of the bond.

5.6 Hybridisation

The treatment of polyatomic molecules in terms of directed orbitals runs into a very serious problem. It is very hard to understand how carbon, with a ground state configuration $1s^2 2s^2 2p^2$, can have a valence of four and have all four bonds equivalent. With two unpaired p electrons, we would expect carbon to be divalent and form compounds like water with a bond angle of approximately 90°. To form four bonds, one of the 2s electrons would need to be promoted to the remaining empty p orbital, giving four unpaired electrons to take part in the formation of chemical bonds. This is not an unreasonable expectation, as the excitation energy required would be less than that gained by bonding. However, it would suggest that carbon should have three bonds arising from the 2p orbitals and another different type of bond arising from interaction with the 2s orbital. In fact, all the bonds in carbon compounds, such as CH_4 and CCl_4, are the same and the compounds have tetrahedral symmetry with bond angles of 109.5°. The dilemma was resolved by Linus Pauling in 1928. He realised that if linear combinations of the 2s and 2p orbitals could form hybrid orbitals, then a higher electron density can be produced between the carbon and the hydrogen or chlorine, producing stronger bonds and more stable molecules (Fig. 5.8). The four hybrid orbitals of carbon are referred to as sp^3 orbitals and, in methane, each can form a σ bond with a 1s orbital of the hydrogen atoms. Bonds formed with hybrid orbitals are highly directional and tend to be stronger than bonds formed with non-hybrid orbitals. A number of types of hybrid orbitals are possible and these are listed in Table 5.4. An indication of how strongly directional a hybrid bond can be was calculated by Pauling from the shapes of orbitals. The factor, given in Table 5.4, though sometimes called bond strength, is not closely related to bond dissociation energies, but it does give some indication of the benefit that can arise from hybridisation.

Hybridisation allows us to understand the geometry of molecules with multiple bonds (Fig. 5.9). In ethane, we have the carbon with sp^3 hybridisation as in methane, and both the carbon–hydrogen and the carbon–carbon bonds are σ bonds. All bond angles are 109°, consistent with the tetrahedral sp^3 hybridisation. In ethene, carbon atoms exhibit sp^2 hybridisation. The carbon–hydrogen bonds are again σ bonds,

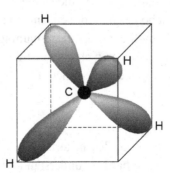

Fig. 5.8. sp^3 hybrid orbitals in methane leading to four equivalent carbon-hydrogen bonds with tetrahedral symmetry and a bond angle of 109.5°.

Table 5.4. Hybridisation and molecular geometry.

Hybrid orbitals	Shape	Bond angle	Pauling factor*	Examples
s			1.00	
p^3		90°	1.73	
sp	linear	180°	1.90	C_2H_2, $BeCl_2$
sp^2	trigonal planar	120°	2.00	BF_3, C_2H_4
sp^3	tetrahedral	109.5°	1.99	CH_4
d^2sp^2	octahedral	90°	2.92	$[Co(NH_3)_6]^{3+}$
dsp^2	tetragonal planar (square)	90°	2.69	$[AuBr_4]^-$

*Pauling estimated the directional specificity of orbitals by calculating the maximum value of the orbital at a given value of r (relative to s = 1.00).

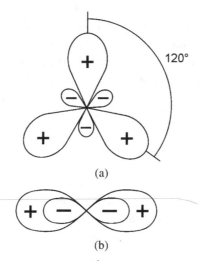

(a)

(b)

Fig. 5.9. Hybrid orbitals: (a) three equivalent sp^2 orbitals oriented as 120° to each other in the plane of the paper, which are characteristic of the orbitals in ethene (ethylene); (b) two equivalent sp orbitals as occur in ethyne (acetylene).

but, in addition to one σ bond linking the carbon atoms, there is a π bond, formed from the remaining p orbitals perpendicular to the molecular plane, leading to the double bond between the atoms. The molecule is planar and the H-C-H bond angle is 121.5° and the H-C-C bond angle is 117°, both very close to the sp^2 hybridisation angle of 120°. In ethyne, the bonding is sp with one σ carbon–carbon bond and two π bonds, leading to a triple bond between the carbon atoms. The molecule is linear, as would be expected for sp hybridisation.

We must not attach too much physical meaning to hybridisation. The electrons in the molecules distribute themselves so as to find the position of minimum energy consistent with the rules of quantum mechanics. The complication arises because we want to describe the resulting wave functions in terms of the atomic orbitals of the atoms which are combining. To put it somewhat frivolously, the electrons themselves go about their work of finding a configuration of minimum energy quite unaware that their original homes were, say, a 1s or 2p atomic orbital. This reverence for their ancestry is an entirely human concept!

5.7 Delocalised orbitals

In order to explain the behaviour of a number of polyatomic molecules with alternating single and double bonds, such as benzene, which exhibit unusual stability, we have seen that the Lewis model found it necessary to introduce a new concept called *resonance*. This assumed that two or more separate structures rapidly alternated. This was an entirely unsatisfactory concept and it is one of the successes of molecular orbital theory that it is able to provide a much more adequate explanation in terms of delocalised orbitals. In order to investigate the molecular orbital approach, which was proposed by Erich Hückel in 1931, we take butadiene as an example. This would be represented in the Lewis model by alternating double and single bonds:

$$H_2C = CH - CH = CH_2.$$

The starting point is to assume that the basic structure of the molecule is provided by the σ bonds and that the π orbitals are assumed to be spread equally over all four carbon atoms (Fig. 5.10). We can write down the secular determinant for this system as we did for the hydrogen molecule.

$$\begin{vmatrix} H_{11} - ES_{11} & H_{12} - ES_{12} & H_{13} - ES_{13} & H_{14} - ES_{14} \\ H_{21} - ES_{21} & H_{22} - ES_{22} & H_{23} - ES_{23} & H_{24} - ES_{24} \\ H_{31} - ES_{31} & H_{32} - ES_{32} & H_{33} - ES_{33} & H_{34} - ES_{34} \\ H_{41} - ES_{41} & H_{42} - ES_{42} & H_{43} - ES_{43} & H_{44} - ES_{44} \end{vmatrix} = 0.$$

To evaluate the determinant in order to calculate the molecular orbital energy level diagram, we make a number of assumptions, some of which we made when studying the hydrogen molecule. All overlap integrals, $S_{ij}(i \neq j)$ are set to zero, and all resonance integrals, H_{ij}, involving neighbouring molecules are represented as β. Other resonance integrals are set to zero. The Coulomb integral, H_{ii}, is written as α. These simplifications enable the secular determinant to be written with all diagonal elements $(\alpha - E)$, all off-diagonal elements between adjacent atoms, β, and all

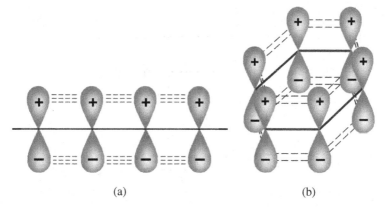

(a)   (b)

Fig. 5.10. Conjugated systems of π orbitals (only π orbitals are illustrated): (a) butadiene; (b) benzene.

other elements equal to zero. The shape of the molecule as formed by the σ bonds determines which atoms are adjacent.

For butadiene (which is linear), we obtain

$$
\begin{vmatrix}
\alpha - E & \beta & 0 & 0 \\
\beta & \alpha - E & \beta & 0 \\
0 & \beta & \alpha - E & \beta \\
0 & 0 & \beta & \alpha - E
\end{vmatrix} = 0.
$$

The equation involving the secular determinant can be reduced to a quartic equation

$$(\alpha - E)^4 - 3(\alpha - E)^2\beta^2 + \beta^4 = 0 \quad \text{and, writing } (\alpha - E)/\beta = x, \text{ we obtain}$$

$x^4 - 3x^2 + 1 = 0$. The roots are $x = \pm 1.62$ and ± 0.62, giving the energies of the four resulting molecular orbitals as:

$$E = \alpha \pm 1.62\beta \quad \text{and} \quad E = \alpha \pm 0.62\beta.$$

The four π electrons fill the lowest two molecular orbitals (which are the bonding orbitals; see Fig. 5.11), giving a total energy for the π electron system as

$$E_\pi = 4\alpha + 4.47\beta.$$

We can estimate the additional energy that arises from the delocalisation of the π electrons (originally termed the resonance energy of the molecule) by computing the energy the molecule would have if the π electrons were localised into the 1,2 and 3,4 double bonds, making β_{23} zero. Solving the modified determinant gives the

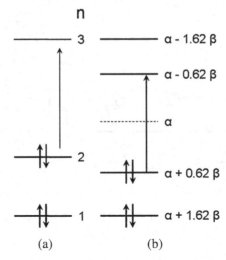

Fig. 5.11. The energy levels of butadiene: (a) calculated using the free electron model, $E = 1157\,\text{kJ mol}^{-1}$, and $\Delta E = h^2/8ma^2\,(3^2 - 2^2)$ giving $\Delta E = 578\,\text{kJ mol}^{-1}$; (b) calculated using Hückel theory, $E = 4\alpha + 4.47\beta$, $\Delta E = E_2 - E_3 = 1.24\beta$, where E is the total energy of the π electron system, and ΔE the energy required to promote an electron from the highest filled energy level to the next available level.

π electron energies in terms of two doubly degenerate energy levels

$$E = \alpha \pm \beta,$$

leading to a total energy for the π electron system

$$E_\pi = 4\alpha + 4\beta.$$

This indicates that the extra energy due to delocalisation is $0.47\,\beta$. We can attempt to evaluate the energy by using estimates obtained by carrying out a similar calculation for benzene. This gives the delocalisation (resonance) energy as 2β, which can be related to a value of β of $-75\,\text{kJ mol}^{-1}$ estimated on the basis of experimental measurements of the stability of a benzene molecule in relation to what might be expected if the double bonds were in fixed positions. This would suggest that the extra stability of butadiene is approximately $36\,\text{kJ mol}^{-1}$. It is probably unwise to attach too much attention to these numerical values (as they lead to a poor account of the spectra), but it is clear that an extra contribution to the stability of the molecule arises naturally from the delocalisation of the electrons.

We can approach this property of delocalisation in a different way by imagining the π electrons to be located in a one-dimensional box (such as we met in Section 3.5) running the length of the molecule. Once we have established the length of the box, a, then we can immediately calculate the energy levels: $E = h^2 n^2 / 8 m a^2$. For a conjugated system with alternating single and double bonds, we may take the average carbon–carbon bond length as 0.14 nm. In the case of butadiene, if we allow for half a bond length at each end of the molecule, we have the total length available to the π electrons $a = 4 \times 0.14$ nm $= 0.56$ nm. Substituting into the equation for energy we find that the energy of an electron in the lowest energy level is 116 kJ mol^{-1} and the energy of the molecule when the first two energy levels are filled by the four π electrons is 1157 kJ mol^{-1} (Fig. 5.11). We can see that when an electron has more space and is more delocalised, its energy will be lower. The energy of delocalisation arises naturally from the calculation with no need for further assumptions.

We can check the accuracy of these free electron calculations by calculating the energy required to promote an electron from the highest filled energy level to the next highest and comparing this with spectroscopic measurements. In the case of butadiene, this energy is 578 kJ mol^{-1}, which would predict a maximum absorption at a wavelength of 207 nm, comparing very favourably with the experimentally-observed spectral absorption at 210 nm. The agreement is better than might be expected and when we calculate the spectral transitions for linear conjugated molecules with six and eight carbon atoms, we find that the calculated wavelengths, 332 nm and 460 nm, are appreciably larger than the corresponding experimental measurements of 247 nm and 286 nm. This gives some indication of the limitations of the free electron model. However, in many dyestuffs, the colour arises from strong transitions of π electrons and more refined versions of the free electron model have been used to calculate the wavelength and the strength of the spectral bands in long-chain polymethylene dyes.

5.8 *Ab initio* calculations

Apart from the highly simplified calculations that we have dealt with above, modern computers provide another route to evaluating molecular properties. These calculations are often referred to as *ab initio* calculations, because they avoid the drastic simplifications made in the previous chapters. This, however, does not mean that they are free from assumptions. In particular, one approximation which is almost universal in molecular quantum mechanics is the *orbital approximation*, which assumes that each electron has its own orbital or wave function. The next step in molecular computation is to generate molecular orbitals. This is usually done by taking a linear combination of atomic orbitals, the LCAO approximation. The atomic orbitals that are used to construct the molecular orbital are referred to as the *basis set*. The atomic orbitals that are employed are often themselves simplifications of the hydrogen-like atomic orbitals. Slater-type orbitals (STOs), which we met in the previous chapter, provide a widely-used approximation. If we express the hydrogen-like wave function as

$$\Psi_{n,l,m_l}(r, \theta, \varphi) = R_{n,l}(r)Y_l^{m_l}(\theta, \varphi)$$

then the angular spherical harmonic component, $Y_l^{m_l}$, is usually common to all approximations. Slater wrote the wave function with a simplified radial component $R_{n,l}(r) = N_{n,l}r^{n-1}\exp(-\zeta r)$, where ζ is the orbital exponent, $\zeta = (Z_{\text{eff}}/n)$, and $N_{n,l}$ is a normalisation constant which ensures that $\int \psi\psi^*\mathrm{d}\tau = 1$ (and the probability of the electron being somewhere is unity). STOs, unlike the hydrogen-like orbitals, do not have radial nodes. Though widely employed, the direct use of STOs has fallen out of favour because multicentre integrals of this form are difficult to calculate. They are usually replaced by Gaussian-type orbitals (GTOs) for which

$$R_n(r) = N_n r^{n-1} \exp(-\alpha r^2).$$

These functions have the invaluable property that the product of two Gaussian functions centred on two different points is itself a Gaussian function centred somewhere on the line joining the two points. This proves a great simplification when evaluating multicentre integrals. Unfortunately, Gaussian integrals do not represent STOs well at short range. To solve this problem, a linear combination of Gaussian functions is used to fit the Slater functions. If three Gaussian functions are used, the resulting orbital is referred to as an STO-3G basis set. A further refinement is to represent the atomic orbitals of valence electrons by the sum of two Slater-type orbitals with different orbital exponents. Basis sets formed in this way are called *double-zeta* basis sets. Each of the STOs is again formed from a linear combination of Gaussian functions. Though the use of Gaussian functions leads to many more integrals, the ease with which they can be evaluated leads to much more rapid computation. The most accurate results are obtained by *configuration interaction* calculations, which involve mixing in excited states. This enables contributions to the *correlation energy* to be included and leads to more accurate results than the *Hartree–Fock limit* (Table 5.1). We have seen (Section 4.9) that the Hartree–Fock limit is the best result that can be obtained when the electron–electron interactions are averaged over the atomic orbitals. The correlation energy arises because the instantaneous interactions of electrons are not the same as those interactions averaged over orbitals.

Ab initio calculations can now be undertaken by non-theoretical chemists, as standard programs have been made available by those developing these *ab initio* computational techniques. The scope and limitations of these calculations may be indicated by the fact that using, for example, the STO-3G basis set, the bond lengths and bond angles of quite substantial organic molecules can be calculated to within a few percent.

5.9   Summary of key principles

This chapter has been largely concerned with the approximate methods that are used to solve the Schrödinger equation for molecules and, to that extent, the principles involved cannot be said to be fundamental. However, there are a number of important comments that can be made. First, apart from the case of the hydrogen molecular ion, quantum mechanics provides no exact answers. Second, the approximate *semi-empirical methods* developed in dealing with molecular bonding and structure do not, except in the very simplest cases, provide answers of quantitative significance. This situation is ameliorated by the fact that modern *ab initio* techniques of quantum mechanical computation can provide useful structural data and energies for quite complex molecules and their interactions. They can be expected to become ever more effective as computing power increases.

However, the most important function of quantum mechanics to date is that it provides a notation and a framework which is invaluable in the discussion of molecular structure and the marshalling of experimental data.

Problems

(1) The bond energy of molecular nitrogen is $942 \, \text{kJ mol}^{-1}$, that of the chlorine molecule $239 \, \text{kJ mol}^{-1}$ and that of the chlorine molecular ion Cl_2^+ $386 \, \text{kJ mol}^{-1}$. How may these results be interpreted in terms of molecular orbital theory?

(2) A polymethylene dye has a linear skeleton of 14 conjugated carbon atoms. Estimate the wavelength at which it will absorb light. Assume the average carbon–carbon bond length to be 140 pm.

(3) Use the Hückel model to evaluate the energy levels of the hydrogen molecule in terms of α and β.

6
Atomic and Molecular Spectra

The electromagnetic waves which are of interest to chemists range from those with wavelengths of $10\,\mathrm{m}$ down to $10^{-10}\,\mathrm{m}$, from radio waves to gamma rays (short wavelength X-rays). Waves in each of these regions induce specific atomic or molecular processes and, in doing so, provide spectra which are one of our most important sources of information about atomic and molecular structure (Table 6.1). A number of factors determine which transitions between atomic or molecular states can be observed spectroscopically. We need to understand at what frequencies spectral lines can be observed, the factors which govern their intensity and when the transitions that could give rise to spectra are forbidden. A consideration of these topics is an important first step to understanding the appearance of spectra and the elucidation of what information can be obtained by studying them.

6.1 Spectroscopy

The experimental techniques used to obtain spectra depend on the wavelength of the electromagnetic radiation employed and will only be described here in a very general way. Spectra can be obtained either by studying the radiation that is absorbed or by studying the emission of radiation from a sample that has been excited to higher energy states. In absorption spectroscopy, the sample is exposed to radiation and that radiation which passes through is examined to identify the missing regions of the spectrum that have been absorbed. When monochromatic radiation is passed through a homogeneous absorbing substance, each layer of thickness $\mathrm{d}x$ absorbs a constant fraction of the incident radiation. The change in the intensity, $\mathrm{d}I$, of the light passing through each layer is $\mathrm{d}I/I = -\alpha c \mathrm{d}x$ where α is a constant and c is the molar concentration of the absorbing substance. Integrating this expression, we obtain $\ln(I/I_0) = -\alpha c l$, where l is the path length of the radiation through the sample. It is more usual to express the absorption using logarithms to base 10 giving $\log(I/I_0) = -\varepsilon c l$, where $\varepsilon = \alpha/2.303$. This relationship is called the *Beer–Lambert law*. ε is the *molar absorption coefficient* (frequently called the *extinction coefficient*). It varies with the wavelength of the radiation, the temperature, and if

Table 6.1. Electromagnetic radiation.

Approximate energies and frequencies of electromagnetic radiation. $Hz = s^{-1}$

Radiation	Energy/kJ mol^{-1}	Frequency/Hz	Wave number/cm^{-1}	Associated with changes in:
γ-rays	10^7	2.5×10^{19}	8.3×10^8	Nuclear configuration
X-rays	10^5	2.5×10^{17}	8.3×10^6	Inner electrons in atoms
Visible and Ultraviolet	10^3	2.5×10^{15}	8.3×10^4	Outer electrons in atoms and molecules
Infrared	10	2.5×10^{13}	8.3×10^2	Molecular vibrations
Microwave	10^{-1}	2.5×10^{11}	8.3	Molecular rotations
Radio	10^{-3}	2.5×10^9	8.3×10^{-2}	Electron spins
	10^{-5}	2.5×10^7	8.3×10^{-4}	Nuclear spins

the absorbing substance is in solution, the solvent. If the concentration is expressed in mol dm^{-3} and the optical path length l in cm, ε has the units dm^3 mol^{-1} cm^{-1}.

The determination of an absorption spectrum requires an appropriate radiation source, a means of focusing the radiation and suitable sample containers that do not themselves absorb significantly in the region of the spectrum under investigation. The radiation which passes through the sample must be analysed as a function of its frequency using a suitable monochromator, such as a prism or grating, which separates radiation of different wavelengths. Finally, the radiation must be detected (Fig. 6.1, Table 6.2). In emission spectra, the sample is excited by exposing it to radiation. The radiation may be absorbed selectively by transitions within the sample and then re-emitted. This re-emitted radiation is analysed in the same way as for absorption spectroscopy.

A fundamentally different technique which is widely used in modern spectroscopy is Fourier transform spectroscopy. In this technique, the beam from a suitable source, containing radiation of a wide range of frequencies, is split into two and remixed after one of the split beams is exposed to a path difference generated by a moving mirror (Fig. 6.1). This results in interference between the two beams. If a sample is placed in the path of the two interfering beams, the path difference between the two beams changes and the interference pattern is modified. The distortion of the pattern, as a function of the changing path length generated by the moving mirror, is recorded. By analysing the interference pattern using a mathematical technique, the Fourier transform, the spectrum can be recovered. The technique has

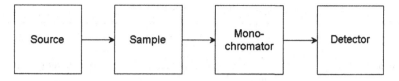

(a) Absorption spectroscopy

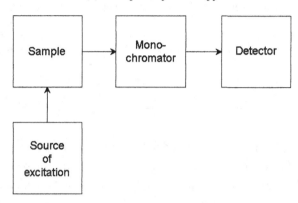

(b) Emission spectroscopy and Raman spectroscopy

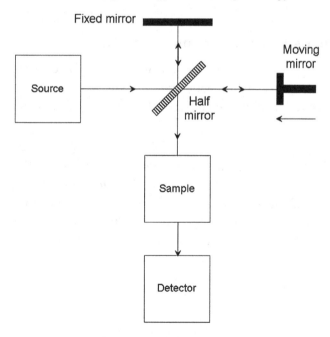

(c) Fourier transform spectroscopy

Fig. 6.1. Spectroscopic techniques (schematic).

Table 6.2. Spectroscopic techniques.

Spectra	Source/Exciter	Monochromator	Detector
Ultraviolet and visible	tungsten, hydrogen or xenon discharge lamps	quartz or alkali fluoride prisms, reflection gratings	photomultiplier tubes or photographic plates
Infrared	heated filament	alkali halide prisms; gratings	thermal detectors; photoconductors
Microwave	Klystron oscillator	electronic modulation of the Klystron	crystal detector; high-frequency radio receiver
Raman	lasers	quartz prism or grating	photoelectron detector or photographic plate

two advantages. First, all wavelengths can be studied simultaneously and, second, by collecting the data over a number of scans, the sensitivity can be much enhanced. Furthermore, it does not exhibit the problems that can occur with prisms and gratings, which lose accuracy as the angle through which the incident beam is deflected increases.

6.2 The intensities of spectroscopic lines

If a molecule with two available states, a and b, is exposed to electromagnetic radiation, it can be stimulated to undergo transitions between the two states (Fig. 6.2). The rate of transition from an excited state a to state b is given by

$$B_{ab} N_a \rho(\nu_{ab}),$$

where $\rho(\nu_{ab})$ is the energy density of radiation of frequency ν_{ab}, N_a is the number of molecules in state a and B_{ab} is known as *Einstein's coefficient for stimulated emission*. (The coefficient of stimulated absorption, B_{ba}, is equal in magnitude to the coefficient of stimulated emission.) Apart from the transitions that are stimulated

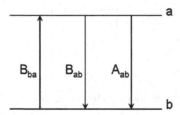

Fig. 6.2. Transitions between two states. The transitions labelled B_{ab} and B_{ba} represent stimulated emission and stimulated absorption, respectively. The transition labelled A_{ab} represents spontaneous emission.

by the incident radiation, excited states can also spontaneously emit photons and return to a lower state. The rate of this process is governed by *Einstein's coefficient for spontaneous emission*, A_{ab}. It can be shown that, for the energy levels in thermal equilibrium, the ratio A_{ab}/B_{ab} is given by

$$\frac{A_{ab}}{B_{ab}} = \frac{8\pi h v_{ab}^3}{c^3}.$$

Thus, the rate of spontaneous transitions is dependent on the cube of the frequency. It is significant for high-energy transitions such as those found in electronic spectra, but negligible for very low-energy transitions involving, for instance, nuclear spins. This process sets the limit on the lifetimes of excited states. Typically, an excited electronic state might have a lifetime of only 10^{-8} s, whereas, for nuclear spins, the lifetime of excited states, if determined only by the rate of spontaneous emission, would be on the order of 10^{19} s (essentially infinite from a practical point of view). These lifetimes are sometimes referred to as the *natural lifetimes* since, in practice, other factors usually shorten the lifetime of excited states.

To evaluate B_{ab}, which determines the intensity of spectroscopic transitions, we must understand how electromagnetic radiation can interact with atomic and molecular systems. We must first consider the nature of light waves. A beam of light generates electrical and magnetic fields at right angles to each other oscillating at the frequency of the radiation (Section 1.8). In the formation of spectra, the oscillating electric field is by far the most significant factor. We have established (Section 2.10) that, when an electrical field interacts with an array of charges, the potential energy arising from the interaction, V, is related to the dipole moment of the charge array, $\mu = \Sigma er$, and the field, $(d\Phi/dr)$, which is the derivative of the electrostatic potential

$$V = -\mu\left(\frac{d\Phi}{dr}\right).$$

Thus, the most important property which determines the interaction of light with atoms or molecules is the dipole moment. The basic rule is: *if the transition between energy levels of an atomic or molecular system is to absorb or emit light, the transition must involve a change in dipole moment.* For a transition between state a and state b, the probability that light will be absorbed or emitted is related to the Einstein coefficient of stimulated emission, B_{ab}, which, in turn, is proportional to the square of a *transition dipole moment*, R_{ab}, which can be expressed as

$$R_{ab} = \int \psi_a \mu \psi_b d\tau.$$

If R_{ab} is nonzero, the transition is said to be spectroscopically *allowed*. If not, it is said to be *forbidden*. It is possible to calculate, from the knowledge of the electronic and nuclear configurations in the initial and final states, the probability of a transition by using the above equation. In practice, it is often more convenient to establish the nonzero values of R_{ab} from a consideration of the symmetry of the initial and final wave functions. An integral of the type $\int \psi_a \mu \psi_b d\tau$, with symmetric limits of integration (such as when the upper limit is a and the lower limit is $-a$), must be symmetric with respect to inversion when the coordinates are reflected through a point centre of symmetry. The dipole moment, μ, always changes sign on inversion. Therefore, the two wave functions must be of different symmetry if the integral itself is to be symmetric. This enables us to determine which transitions will be observed spectroscopically in atoms or molecules which have a centre of inversion. If the wave function of one state is even and the other odd, the integral will be unchanged on inversion and the transition will be permitted. Transitions are only allowed when there is a change of symmetry from odd to even or even to odd. In atoms, the wave functions are even or odd depending on the value of the angular momentum quantum number, l. s and d states are even and p and f states are odd. Transitions from s $\rightarrow$ p and p $\rightarrow$ d are observed spectroscopically, but those between states of the same symmetry are not. In a similar manner, in molecules with a centre of inversion, transitions are allowed between states of different symmetry. In the notation introduced in Section 5.4, transitions between gerade and ungerade states are spectroscopically allowed but those between states of the same symmetry are not. Such considerations of the symmetry of the wave functions of the states involved enable us to establish sets of conditions that determine whether spectroscopic transitions are allowed. Such conditions are called *selection rules*. Selection rules play an important part in aiding the interpretation of spectra but it is important to note that there are circumstances where they are not rigorously adhered to. We will examine selection rules in greater detail in the sections dealing with specific spectroscopic techniques.

The transitions we have described are known as *electric dipole transitions* and are by far the most important in determining the spectra of atoms and molecules. However, there are other interactions that can give rise to spectra. One such source of spectra are *magnetic dipole transitions*. The spin magnetic moment of a nucleus or electron can interact with the oscillating magnetic field component of radiation to produce transitions which, though much weaker than electric dipole transitions (on the order of 10^{-5} times weaker), have proved important in chemistry. Another property which can give rise to spectra are the quadrupole moments of molecules, which can interact with electric field gradients to facilitate spectroscopic transitions, but the transitions are even weaker than magnetic dipole transitions.

6.3 Spectroscopic line widths

A number of factors determine the width of spectroscopic lines. The most fundamental is the Heisenberg uncertainty principle in the form

$$\delta E \times \delta t \geq \frac{h}{4\pi} \approx 10^{-34} \, \text{J s}.$$

Expressing this relation in terms of frequencies, as $E = h\nu$, we obtain

$$\delta\nu \approx \frac{\delta E}{h} \approx \frac{1}{(4\pi\delta t)} \approx \frac{1}{(10\,\delta t)}\text{s}^{-1}.$$

If a state is long-lived, δt is large and the uncertainty in the energy will be correspondingly small. This leads to a sharp spectroscopic line. However, if a state is short-lived and δt is small, the uncertainty in the energy δE will be larger, leading to a broad spectroscopic line. A typical excited electronic state will have a lifetime of approximately 10^{-8} s, which leads to an uncertainty in the energy δE on the order of 10^{-26} J and an uncertainty in the frequency $\delta\nu$ on the order of 10^7 Hz. Electronic transitions typically have frequencies on the order of 10^{15} Hz and, despite the comparatively large absolute value of the uncertainty, from a practical point of view, the lines can appear extremely sharp. The maximum lifetime of an excited state is determined by the spontaneous emission process described above, which establishes what is referred to as the *natural lifetime*. However, other factors, such as *collisional deactivation*, when colliding molecules take away the energy of excitation and return an atom or molecule to a state of lower energy, can play an important part in determining the lifetime of an excited state.

The lifetime of an excited state determines, through the uncertainty principle, the limit of the accuracy with which its energy can be defined, but there are other factors which can contribute to the width of spectroscopic lines. They include *collision broadening*, which is caused by the impact of collisions altering the atomic and molecular energy levels. In gases, the collision frequency depends on the pressure and collision broadening can be reduced by making measurements at low pressure. In gases and liquids, Doppler shifts, due to the rapid motion of the molecules, can alter the frequency of the transition that is observed. The frequency is shifted depending on the ratio of the velocity of the atom or molecule to that of the speed of light. Those approaching the detector appear of higher frequency and those moving in the opposite direction of lower frequency. As the motion of the source can involve a distribution of velocities both in magnitude and direction relative to the detector, this can contribute to the broadening of spectroscopic lines.

6.4 Atomic spectra

We have established in Section 4.2 that the three-dimensional solution of the Schrödinger equation for an electron rotating around a central charge involves four quantum numbers; n, the principal quantum number, determines the energy; l, the orbital quantum number, determines the shape of the electron distribution and the magnitude of the electronic angular momentum; and m_l, the magnetic quantum number, determines the direction of the orbital in space and its behaviour in a magnetic field. The fourth quantum number, m_s, determines the direction of the angular momentum of the electron which can take on two orientations represented by $\pm 1/2$. The quantum number l can take on the values $0, 1, \ldots (n - 1)$, and m_l has the values $0, \pm 1, \ldots, \pm l$. When there is only one electron, the extra energy levels corresponding to the orbital and magnetic quantum numbers are not observed because the energy is dependent solely on the principal quantum number, n. The observed energy levels are made up of a number of components of equal energy and are said to be *degenerate*. However, in the presence of other electrons, orbitals with different values of l do not have the same energy (Section 4.4) and this leads to a more complicated spectrum. For single electrons, we represent their orbital angular momentum by the letters s, p, $\ldots$ For atoms with more than one electron we use a similar notation. We define a total orbital angular momentum, L, as the sum of all the individual electronic orbital quantum numbers, l_i. For two electrons, L can take on all the integral values from $(l_1 + l_2)$ to $|l_1 - l_2|$. We also define a total spin quantum number, S, which goes from a maximum, corresponding to all the spins being parallel, in integer steps down to a minimum which is zero for an even number of electrons and a half for an odd number of electrons.

Thus for n electrons each with a spin of 1/2 we have, when n is odd,

$$S = n/2, (n/2 - 1), \ldots, 1/2,$$

and when n is even,

$$S = n/2, (n/2 - 1), \ldots, 0.$$

If we know the values of L and S, we can obtain the quantum numbers corresponding to the total angular momentum of the atom, J, given by

$$J = (L + S), (L + S - 1) \ldots |L - S|.$$

If L is greater than S, the number of allowed values of J is $(2S+1)$. When $L = 0$, we obtain $J = \pm 1/2$ but, as the negative sign of J is not material, this in fact represents only one energy level. States with a value of $(2S+1)$ of one are called *singlet* states, and those states with $(2S + 1)$ equal to two and three are called *doublet* and *triplet* states, respectively.

The two new quantum numbers we have generated, L and S, provide a basis for describing the electronic state of atoms.

The states for which $L = 0, 1, 2, 3, 4, \ldots$ are, by analogy with the notation for individual electrons, designated S, P, D, F and, together with the values for the total spin quantum number and the total angular momentum quantum number, J, used to define what is called a *term symbol*, written

$$^{2S+1}L_J.$$

The superscript gives the *multiplicity* of the state and the subscript indicates the values of the total orbital angular momentum quantum number, J.

Let us apply this notation to simple atoms. In the case of hydrogen, which possesses only one electron, the term symbol for the atom is just that for the single electron. In the ground state, the electron is in the 1s orbital and the atomic term symbol is $^2S_{1/2}$. Hydrogen also has states $^2P_{1/2}, ^2P_{3/2}$ and $^2D_{3/2}, ^2D_{5/2}$ etc. and these are doublet states, though the separation of the levels is very small. This splitting is caused by the magnetic moment of the spinning electron interacting with the magnetic field created by the motion of the electron around the nucleus, an interaction known as *spin-orbit coupling*. The magnetic field generated in hydrogen is very small, but for elements with a larger nuclear charge, where the electrons are further from the nucleus and moving with higher velocity, the splitting can be significant, increasing rapidly with increasing atomic number, Z. The splitting for the ^2P energy levels of Cs is almost 2000 times larger than for the same levels of Li.

Hydrogen is unique in that, for a given value of n, the S, P and D states all have the same energy, leading to a very simple spectrum. The lithium atom has the same states but now, because of the inner shell of 1 s electrons, the energies of the S, P and D states are different because they give rise to different degrees of shielding of the outer electron from the nuclear charge as a result of the different radial distribution of the electron (Section 4.4). For instance, the effective nuclear charge, Z_{eff}, experienced by a 2s electron is greater than that for a 2p electron and, from

$$E = \frac{-R_H Z_{\text{eff}}^2}{n^2},$$

we see that the 2s electron with the larger Z_{eff} is lower in energy (that is, has a larger negative energy) and is more stable than the 2 p electron. Thus, the S states of lithium lie at lower energies than the P states (Fig. 4.4). The separation of the S, P and higher states leads to a more complicated spectrum for lithium and the other alkali metals than is observed for hydrogen. In hydrogen, all possible transitions between the levels are permitted as they are defined only by the principal quantum number n. However, for the alkali metals, though there are more energy levels, not all transitions are spectroscopically active. The transitions that are allowed are determined by *selection rules*.

The symmetry of states for which $L = 0, 1, 2, \ldots$ alternates; those with even values are symmetric and those with odd values are anti-symmetric. The selection rule for orbital angular momentum is

$$\Delta l = \pm 1$$

$$\Delta L = 0, \pm 1, \quad \text{but } L = 0 \rightarrow L = 0 \text{ is not allowed.}$$

The electrostatic field generated by radiation cannot interact with the electron spins and so the selection rule for S is

$$\Delta S = 0$$

and, for J, the total orbital angular momentum quantum number, $\Delta J = 0, \pm 1$ with the additional restriction that a $J = 0$ state cannot undergo a transition to another $J = 0$ state. These selection rules are summarised in Table 6.3.

The selection rules determine the appearance of the spectrum. The allowed transitions for sodium are illustrated in Fig. 6.3. The alkali metal lines are doublets and the intense yellow line that is characteristic of sodium is a doublet which arises from a transition between the $3^2S_{1/2}$ and the $3^2P_{1/2,3/2}$ levels (Fig. 6.4).

6.5 Two-electron spectra

The spectrum of the helium atom with its two electrons provides an interesting example of the application of selection rules. In the ground state, there are two electrons in the 1s orbital and, according to the Pauli exclusion principle, they will have anti-parallel spins. This corresponds to a $^1S_0$ state. However, when we promote one of the electrons to a different orbital, the spins can be either parallel or

Table 6.3. Selection rules for atomic spectra.

For Russell-Saunders coupling $L = \Sigma l_i$, $S = \Sigma s_i$ and $J = L + S$.
The selection rules are:
$\Delta l_i = \pm 1, \quad \Delta L = 0, \pm 1$
but $L = 0 \rightarrow L = 0$ is not allowed.
$\Delta S = 0$
$\Delta J = 0, \pm 1$

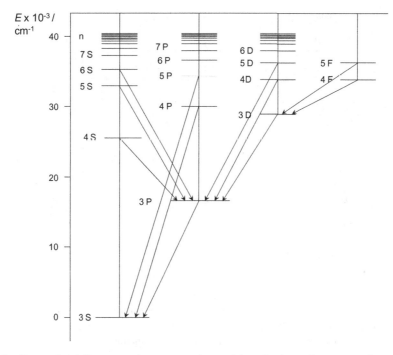

Fig. 6.3. Energy level diagram and spectroscopic transitions for the sodium atom (after Grotrian).

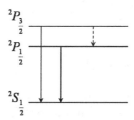

Fig. 6.4. Origin of the doublet structure in the yellow sodium line (not to scale). The doublet line arises from the selection rules $\Delta L = \pm 1$, $\Delta J = 0, \pm 1$. The transition indicated by the broken line is forbidden.

anti-parallel. The anti-parallel spins lead to a further series of singlet states such as $^1P_1$ and $^1D_2$ (arising from the $1s^1 2p^1$ and the $1s^1 3d^1$ configurations, respectively). By contrast, the configurations with parallel spins lead to triplet states of the form $^3S_1$, $^3P_{2,1,0}$, $^3D_{3,2,1}$ (Fig. 6.5). Transitions are allowed between those states for which $\Delta L = \pm 1$ but the spin selection rule, $\Delta S = 0$, forbids any transitions between the singlet levels and triplet levels. This leads to two independent forms of helium which can only inter-convert with the greatest difficulty. The lowest energy level is the singlet level with anti-parallel electron spins and the most common form of helium is that with singlet energy levels. The alternative form with parallel electron spins comprises the triplet states. Similar spectra consisting of two series of states

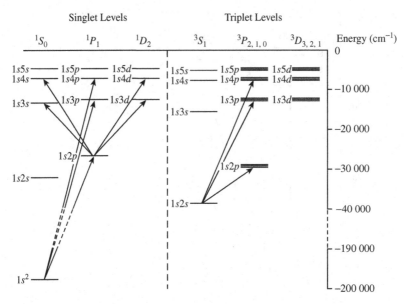

Fig. 6.5. Energy levels of the helium atom showing spectroscopically allowed transitions. Transitions between the singlet and triplet states are forbidden and the spectrum of helium appears to originate from two different species, which were labelled ortho- and para-helium.

which do not readily combine with each other are also characteristic of the alkaline earth elements, which, like helium, have two valence electrons.

6.6 Russell–Saunders coupling

In establishing the notation for the states of atoms, using the term symbols we introduced above, we have made an implicit assumption. We have assumed that the individual electronic angular momenta can add together to produce a resultant total orbital angular momentum for the atom which is itself quantised. This can only happen if the orbital angular momentum of an electron is 'aware' of the behaviour of this property in other electrons. In the same way, we have assumed that the individual electron spins will add together to produce a resultant spin which is itself quantised. For this to happen, the orbital angular momenta must interact, or, to use the term most frequently employed, *couple* with each other to produce a *quantised resultant*. The electron spins must also couple to produce a quantised resultant. When this occurs, it is known as Russell–Saunders coupling or $L–S$ coupling (Fig. 6.6). It turns out to be a very good approximation for atoms with atomic number less than about 40.

Russell–Saunders coupling occurs when the electrostatic interactions of the electrons are dominant. However, in heavier atoms, spin-orbit coupling, the magnetic interaction between the orbital angular momentum and the electron spin becomes significant. In atoms of higher atomic number, the electrons move with higher

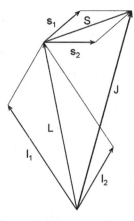

Fig. 6.6. Russell–Saunders coupling.

velocities and consequently generate stronger magnetic fields. This alternative pattern of interaction dominates for atoms of high atomic number and is referred to as *j–j coupling*. In this case, the individual orbital angular momenta interact with the electron spin of the same electron to produce a resultant, the total angular momentum of the electron, which is itself quantized:

$$j_i = l_i + s_i.$$

The total angular momenta of the individual electrons then couple to give a resultant total angular momentum for the atom, which is again quantized:

$$J = j_1 + j_2 + j_3, \ldots.$$

Under these circumstances, L and S are no longer well-defined quantum numbers and this leads to a breakdown of the selection rules based on Russell–Saunders coupling. Thus, in the atomic spectrum of mercury, we find quite intense spectroscopic lines that result from transitions between singlet and triplet states. This shows that, for heavier atoms, the spin selection rule, $\Delta S = 0$, no longer operates as strictly as for lighter atoms.

Nevertheless, it is the practice, even in cases where Russell–Saunders coupling breaks down, to define atomic states using the term symbols that are based on the values of L and S.

6.7 Molecular spectra

The total energy of a molecule is, to a very good approximation, the sum of a number of different types of energy:

$$E_{\text{total}} = E_{\text{elec}} + E_{\text{vib}} + E_{\text{rot}} + E_{\text{trans}}$$

Each contribution gives rise to different regions of the electromagnetic spectrum (Table 6.1). Electronic energy levels are usually separated by energies on the order of 10^3 kJ mol^{-1}, which correspond to photons in the ultraviolet or visible regions of the electromagnetic spectrum. The energy levels that arise from the vibrational motion of the molecules are much closer together, on the order of 10 kJ mol^{-1}. The spectra that result directly from vibrational transitions are observed in the infrared region. Rotational energy levels lie even closer together and can be observed in the microwave region. In fact, it is the rotational transitions in water that absorb the energy in a microwave oven. The vibration and rotational spectra may each be observed directly but, in favourable circumstances, the effect of vibrational and rotational transitions can also be seen as the fine structure of an electronic spectrum. For example, in iodine, the transitions from one electronic state to another are made up of a series of 'bands', each corresponding to a specific change in vibrational energy. Within these vibrational bands, lines arising from the many possible rotational changes can be seen. Thus, each line corresponds to a specific change in electronic energy, a specific change in vibrational energy and a specific change in the rotational energy:

$$\Delta E = \Delta E_{\text{elec}} + \Delta E_{\text{vib}} + \Delta E_{\text{rot}}.$$

In a similar manner, the vibrational transitions observed in infrared spectra are made up of many lines due to the different rotational changes that can accompany the transition. These rotational features can be used to calculate the moment of inertia and hence bond lengths and bond angles of molecules. Different chemical bonds give rise to characteristic vibrational frequencies which can be used to identify chemical structure. Even when a full resolution of the spectrum is not possible, it can be regarded as a fingerprint which, when compared with previously-determined spectra, can identify the chemical structures involved. We will now consider these different aspects of molecular spectra in more detail.

6.8 Rotational spectra

Pure rotational spectra can be used to determine the moments of inertia of molecules and hence bond lengths and structural information. Molecules may have up to three moments of inertia, depending on their shape, and these, in turn, determine the character of the rotational spectra that are observed (Fig. 6.7).

(i) Linear molecules: these have two moments of inertia about two axes at right angles which are equal and a moment of inertia about a third axis which is zero, $I_A = I_B$ and $I_C = 0$. Examples are diatomic molecules and other linear molecules such as HCN.

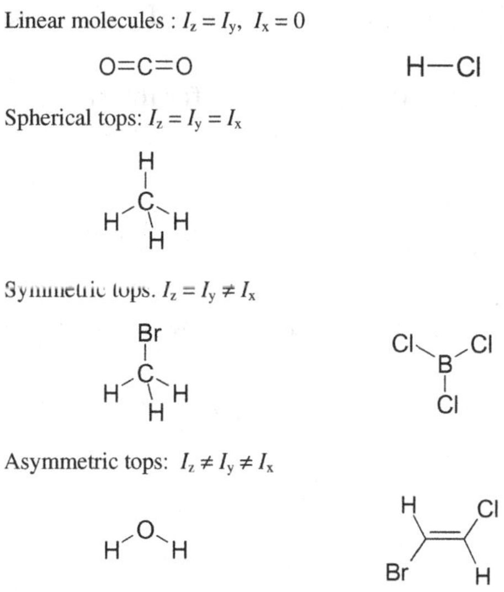

Linear molecules : $I_z = I_y$, $I_x = 0$

O=C=O   H—Cl

Spherical tops: $I_z = I_y = I_x$

Symmetric tops. $I_z = I_y \neq I_x$

Asymmetric tops: $I_z \neq I_y \neq I_x$

Fig. 6.7. Molecular shapes and moments of inertia.

(ii) Symmetric tops: $I_A = I_B$ and $I_C \neq 0$. For example, CH_3F.
(iii) Spherical tops: $I_A = I_B = I_C$. For example, SF_6 and CF_4.
(iv) Asymmetric tops: $I_A \neq I_B \neq I_C$. For example, H_2O.

Each of these characteristic symmetries gives rise to different patterns of rotational spectra. The simplest case, and the one to which we will give most attention, is that of diatomic molecules. The moment of inertia of a diatomic molecule is given by (Section 3.6)

$$I = \Sigma m_i r_i^2,$$

which for a homonuclear diatomic molecule gives

$$I = \left(\frac{1}{2}\right) mR^2,$$

where m is the atomic mass and R the bond length.

However, for a diatomic molecule to have a rotational spectrum, it must have a permanent dipole moment which, when the molecule rotates, can interact with the electromagnetic radiation. Clearly, homonuclear molecules do not satisfy this criterion and are not active in the microwave region. Heteronuclear diatomic molecules, on the other hand, do give rise to a microwave spectrum. For a heteronuclear diatomic molecule, we obtain, by summing the product of each of the atomic masses and the square of their distances from the centre of gravity of the molecule, $I = m_1 m_2/(m_1 + m_2)R^2$, often written as $I = \mu R^2$, where μ is the reduced mass of the molecule (not to be confused with the same symbol, μ, used for dipole moments).

The solution of the Schrödinger equation for the energy levels of a rigid rotor (Section 3.6) gives

$$E_J = \left(\frac{h^2}{8\pi^2 I}\right) J(J+1), \quad \text{where } J = 0, 1, 2, 3, \ldots.$$

This is often written in terms of wave numbers with the units cm^{-1} as

$$\bar{v}_J = \frac{E_J}{hc} = BJ(J+1),$$

where B, the rotational constant, is given by

$$B = \left(\frac{h}{8\pi^2 Ic}\right).$$

This gives a series of energy levels $0, 2B, 6B, 12B, 20B, \ldots$ (Fig. 6.8) with the spacing between them given by the equation

$$\bar{v} = \bar{v}_{J+1} - \bar{v}_J = 2B(J+1) \quad \text{and} \quad \Delta\bar{v} = 2B.$$

The symmetry of the rotational energy levels alternates and the selection rule is $\Delta J = \pm 1$, leading to a series of equally-spaced lines at $2B, 4B, 6B, \ldots$. Thus, the lines in a pure rotational spectrum are approximately $2B$ apart.

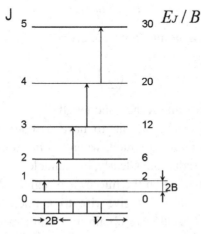

Fig. 6.8. Energy levels of the rigid rotor showing the allowed spectroscopic transitions, which lead to equally spaced lines separated by $2B$.

Example: the microwave spectrum of HF shows a series of lines approximately $41 \, cm^{-1}$ apart. We can use this data to calculate the bond length in HF:

$$\Delta \bar{v} = 2B = 41 \, cm^{-1}, \quad \text{giving } B = 20.5 \times 100 \, m^{-1}.$$

$$I = \frac{h}{8\pi^2 Bc} = \frac{6.63 \times 10^{-34}}{8\pi^2 \times 3 \times 10^8 \times 20.5 \times 100} = 1.37 \times 10^{-47} \, kg \, m^2$$

$$R^2 = \frac{I}{\mu},$$

where the reduced mass

$$\mu = \frac{m_1 m_2}{(m_1 + m_2)}$$

$$\mu = \frac{1 \times 19 \times 1.66 . 10^{-27}}{(1 + 19)} \, kg$$

$$R^2 = \frac{I}{\mu} = \frac{1.37 \times 10^{-47}}{1.58 \times 10^{-27}} = 0.87 \times 10^{-20} \, m^2$$

$$R = 0.93 \times 10^{-10} \, m = 93 \, pm.$$

The most accurate experimentally determined value is 92 pm.

The intensity of rotational transitions goes through a maximum as J increases. This arises from two competing factors. In the first place, the degeneracy of the rotational levels increases as $(2J + 1)$ since the angular momentum may take on quantised values from $-J$ through zero to $+J$. This would suggest an increasing intensity as J rises. This, however, is compensated by the fact that the Boltzmann factor (explained in Section 7.5) decreases with increasing J as the energy increases. Combining the two factors, we obtain

$$\frac{N_J}{N_0} = (2J + 1) \exp\left[\frac{-(Bhc \, J(J + 1))}{kT}\right].$$

The population of rotational levels is particularly sensitive to temperature, because, unlike translational levels, which are extremely close together compared with kT, and vibrational levels, which are very far apart compared with kT, rotational levels have energies very close to the value of kT at room temperature. The two competing factors of degeneracy and the Boltzmann factor lead to a maximum in the intensity of a series of rotational lines at a value of J where the population is maximum, and $(dN_J/dJ) = 0$, giving

$$J_{max} = \left(\frac{kT}{2hcB}\right)^{\frac{1}{2}} - \frac{1}{2}.$$

Though the main features of the rotational spectra of diatomic molecules (see Fig. 6.12) can be interpreted in terms of the rigid rotor model, as the molecules rotate faster, the bond between the atoms stretches, increasing the moment of inertia. This small but significant effect causes the lines to become slightly closer together as J increases. Linear molecules with a permanent dipole moment give rise to similar spectra to diatomics, but the moment of inertia is often greater and the lines appear closer together.

Non-linear molecules give rise to much more complicated spectra. For rigid symmetric tops, we define a quantum number J to represent the total angular momentum and a second quantum number K to represent the projection of the angular momentum about the principal axis of symmetry. The spectrum is independent of the second quantum number K and rotation about the principal axis of symmetry does not contribute to the rotational spectrum. However, if the centrifugal stretching takes place, the spectrum becomes much more complicated and the value of K leads to the splitting of the rotational lines. With this extra information, it is possible for the bond lengths and the geometry of a symmetrical top molecule to be determined to high accuracy.

Spherical tops do not give rise to pure rotational spectra, but sometimes weak lines can be detected due to centrifugal distortion. Asymmetric tops have very complicated rotational spectra which defy simple analysis and are beyond the scope of this treatment.

6.9 Vibrational spectra

The simplest case we can consider is that of a diatomic molecule. If it behaves as a harmonic oscillator, the Schrödinger equation tells us that the energy levels are equally spaced (Section 3.7; Fig. 6.9):

$$E_v = \left(v + \frac{1}{2}\right)h\nu,$$

where v is the vibrational quantum number. The lowest energy level, when $v = 0$, is $E_0 = \frac{1}{2}h\nu$. This is the *zero-point energy*. The selection rule for a harmonic oscillator is $\Delta v = \pm 1$, leading to a single spectral line with a frequency given by $\Delta E = h\nu$.

However, for real molecules, the harmonic oscillator model breaks down. The potential energy curves are similar to that shown in Fig. 6.10 and the energy levels move closer together with increasing values of the vibrational quantum number, v. Eventually, the separation of the levels falls to zero and, at this point, the molecule dissociates into atoms. We can allow for this by writing the vibrational energy as

$$\bar{\nu}_v = \frac{E_v}{hc} = \left(v + \frac{1}{2}\right)\omega_e - \left(v + \frac{1}{2}\right)^2 \omega_e x_e,$$

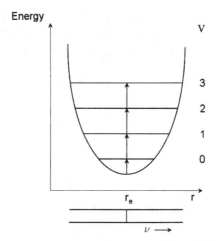

Fig. 6.9. Vibrational energy levels for a harmonic oscillator. The allowed transitions give rise to a single spectral line.

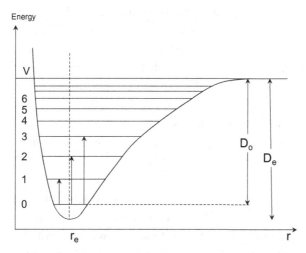

Fig. 6.10. The potential energy function for a diatomic molecule. The energy levels get closer together as the vibrational quantum number increases until dissociation occurs. Most transitions originate from the lowest vibrational level. D_0 is the dissociation energy. D_e is the depth of the potential energy well. D_e is greater than the dissociation energy by the zero point energy, $\frac{1}{2}h\nu$.

where the term $\omega_e x_e$ represents the contribution of *anharmonicity* and ω_e is the vibrational frequency, expressed in wave numbers, extrapolated to $v = 0$. For anharmonic oscillators, the vibrational selection rule becomes less stringent:

$$\Delta v = \pm 1, \pm 2, \pm 3, \ldots .$$

However, the line given by $\Delta v = \pm 1$ is the most intense. Unlike rotational energy levels, vibrational energy levels are widely separated compared with the thermal energy kT and so the overwhelming majority of molecules occupy the lowest vibrational level. This, in conjunction with the selection rule, leads to a series of lines corresponding to transitions from the ground state to level v', given by

$$\bar{v} = v'\omega_e[1 - (v' + 1)x_e]$$

each separated, very approximately, by ω_e (if $\omega_e x_e$ is small). The first line, at ω_e, is called the *fundamental* line and the others, which are usually very much weaker, the *overtones*.

Vibrational spectra are not observed in homonuclear diatomic molecules, as the vibrational motion does not cause a change in dipole moment. However, such spectra can be observed in heteronuclear diatomics.

6.10 Vibrational-rotational spectra

Vibrating diatomic molecules also rotate at the same time, giving rise to a complex structure associated with each vibrational transition. The rotational lines that accompany the vibrational transition arise in a similar manner to those of pure rotational spectra. However, the change in the moment of inertia that accompanies the vibrational stretching leads to a somewhat different appearance. B, the parameter which determines the separation of rotational lines, is inversely proportional to the moment of inertia, which in turn is proportional to the square of the inter-nuclear distance, r. Thus, when the bond length is larger, B becomes smaller.

The total energy of a rotating vibrating molecule can be written as

$$E_{total} = E_{vib} + E_{rot}.$$

The rotational energy contributes only a very small fraction to the total energy, but it can lead to a distinct pattern of fine structure in the vibrational spectrum. Ignoring the small term representing centrifugal stretching, we may write the total energy in terms of the quantum numbers for rotational and vibration, J and v, expressed as wave numbers:

$$\bar{v}_{J,v} = \frac{E_{J,v}}{hc} = BJ(J + 1) + \left(v + \frac{1}{2}\right)\omega_e - \left(v + \frac{1}{2}\right)^2 \omega_e x_e$$

The selection rules are the same as for separate vibrational and rotational transitions,

$$\Delta v = \pm 1, \pm 2, \pm 3, \ldots \quad \text{and} \quad \Delta J = \pm 1$$

for diatomic molecules. ($\Delta J = 0$ is allowed only for molecules in which the electrons have angular momentum about the inter-nuclear axis (see Section 6.14) of which nitric oxide is the only example among stable diatomic molecules.) Considering the fundamental transition from $v = 0$ to $v = 1$ and substituting

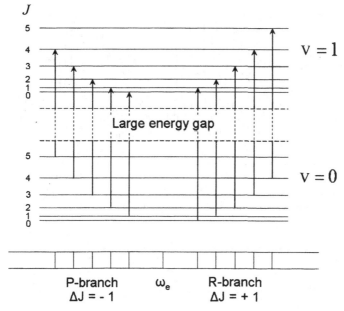

Fig. 6.11. Vibrational-rotational levels of a diatomic molecule exhibiting simple harmonic oscillation and behaving as a rigid rotor. The resulting spectral transitions give rise to a series of equally spaced lines either side of the band origin which corresponds to the $J = 0 \rightarrow J = 0$ transition, which is forbidden by the selection rules.

the quantum numbers consistent with the selection rules into the energy expression, we obtain

$$\bar{v} = \omega_e(1 - 2x_e) \pm 2B(J + 1) = \omega_0 \pm 2B(J + 1),$$

where ω_0 is the frequency that would occur if a vibrational transition with no rotational changes (for which $\Delta J = 0$) were permitted. It is identified as the *band origin* (Fig. 6.11). If $\Delta J = +1$, we obtain a positive sign in the above expression, and $\Delta J = -1$ corresponds to the negative sign. The spectrum consists of equally spaced lines on either side of a gap where the $\Delta J = 0$ line would appear if it were allowed by the selection rules. Lines to the low-frequency (low-energy) side of this gap, for which $\Delta J = -1$, are called the *P branch*, and those on the high-frequency side, corresponding to $\Delta J = +1$, are called the *R branch* (Fig. 6.12). In circumstances when $\Delta J = 0$ is allowed, a central compact series of lines, called a Q branch, is observed (Fig. 6.13).

We have made the assumption that the moment of inertia, and hence B, does not depend on the rotational quantum number. In fact, this turns out to be a very reasonable assumption in view of the very small magnitude of the rotational energy contributions. A more significant contribution to the change in the moment of inertia is due to the anharmonic nature of molecular vibrations. The bond lengths increase

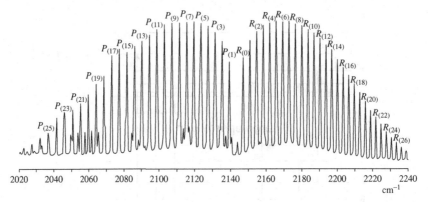

Fig. 6.12. P and R branches in the infrared spectrum of carbon monoxide. (The weaker lines are due to ^{13}CO.)

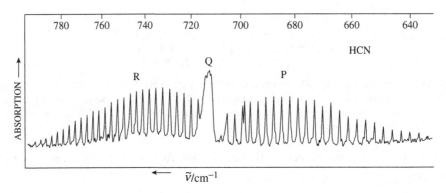

Fig. 6.13. The infrared spectrum of hydrogen cyanide (HCN) showing a Q branch.

as the vibrational energy increases, leading to an increase in the moment of inertia and a decrease in B, which we can express as

$$B_0 = B_v - \alpha \left(v + \frac{1}{2} \right)$$

This effect is relatively small for low values of the vibrational quantum number, but becomes more significant for transitions from the ground state to high values of v.

For molecules with a centre of symmetry, the intensity of rotational lines can be affected by the spins of the component nuclei. Nuclear spin wave functions can be either symmetric or anti-symmetric, and rotational levels alternate in their symmetry. For nuclei of odd mass numbers (which are fermions), the overall wave function must be anti-symmetric, and only alternate rotational levels are compatible with any given nuclear spin symmetry. In the case of hydrogen, the two nuclei, each with nuclear spin $I = 1/2$, can either have parallel spins ($I = 1$), or anti-parallel spins

($I = 0$). For molecules with parallel nuclear spins, only anti-symmetric rotational levels with odd values of J are consistent with the overall symmetry requirement. As the degeneracies, and hence the statistical weights of the levels, are given by ($2I+1$), lines arising from levels for which J is odd have three times the intensity of those arising from even levels. This gives rise to spectral lines of alternating intensity which can be observed only in the Raman spectra (Section 6.13), as hydrogen, which is a homonuclear diatomic molecule, is inactive in the infrared. As interchange of nuclear spins is not easy, hydrogen exists in two forms: the *ortho*-form with parallel nuclear spins and the *para*-form with opposed nuclear spins. The ortho-form, with a statistical weight of three, comprises 75% of hydrogen gas at normal temperatures. In carbon dioxide, where the nuclei have zero spin and the overall wave function is symmetric, only even values of J are acceptable. All the lines arising from the lower states for which J is odd are absent and only alternate lines appear in the Raman spectrum.

6.11 Vibrations of polyatomic molecules

As might be expected, when we go from diatomic molecules to polyatomic molecules, the spectra become much more complex. The number of vibrations of a polyatomic molecule can be defined in terms of the number of coordinates which are required to specify the positions of all the atoms (Section 2.3). This is given by $3N$, where N is the number of atoms. However, some of these coordinates are used to specify types of motion other than vibrations. The translational motion of the molecule in three dimensions can be represented using three coordinates and, in general, the rotational motion of the molecule about three axes requires another three. The remaining coordinates, $3N - 6$, are associated with the vibrations. If the molecule is linear, its rotation can be described in terms of just two coordinates, and the number of vibrational coordinates becomes $3N - 5$. For a diatomic molecule for which $N = 2$, we have just one vibration, whereas in a non-linear triatomic molecule, such as water, we have three vibrational modes.

For a polyatomic molecule of N atoms, there are $N(N - 1)/2$ inter-nuclear distances and, for a non-linear molecule, only $3N - 6$ independent vibrational coordinates. For example, in methane, with five atoms, there are ten inter-nuclear distances, but only nine vibrational modes. There is therefore a restriction on how we can define the vibrations of the molecule. We find that for each molecule we can identify vibrational motions that are known as *normal modes*. These are vibrations in which all the atoms move with the same frequency and each normal mode can be excited separately. If we know the symmetry of the molecule, it is possible to identify these normal modes. Vibrations can be identified as bond stretching, bond bending or bond twisting. If the vibration is associated with a change of dipole

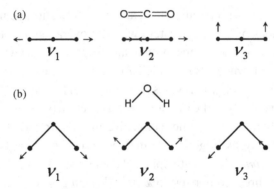

Fig. 6.14. (a) The vibrational modes of a linear molecule, such as carbon dioxide. v_1 corresponds to symmetric stretching, v_2 to anti-symmetric stretching, and v_3 to bending (doubly degenerate); (b) the vibrational modes of a non-linear molecule, such as water. v_1 corresponds to symmetric stretching, v_2 to symmetric bending and v_3 to anti-symmetric stretching. v_1 and v_2 are parallel vibrations and v_3 is a perpendicular vibration.

moment, it will give rise to a spectrum in the infrared region. If there is no change of dipole moment, it will be inactive in the infrared. Carbon dioxide, a linear molecule with three atoms, has four normal modes of vibration, illustrated in Fig. 6.14. The symmetrical stretching clearly will not result in a change of dipole moment and is therefore inactive, whereas the asymmetric stretching motion will give rise to a transition observable in the infrared spectrum. The two identical bending modes, which are perpendicular to each other, are also active in the infrared. A non-linear triatomic molecule such as water will have three normal vibrational modes, as illustrated in Fig. 6.14. Thus, the symmetric stretching, symmetric bending and anti-symmetric stretching modes all result in changes in the dipole moment and are active in the infrared. For the first two, the change of dipole moment lies along the axis of symmetry, whereas the anti-symmetric stretching mode causes a change in dipole moment perpendicular to that axis. Such vibrational modes are termed *parallel* and *perpendicular*, respectively. This distinction is particularly significant when we consider the rotational structure accompanying the vibrational transitions. For linear molecules, parallel vibrations show much the same rotational structure as we see in diatomic molecules, which is not surprising as the motion of the atoms is similar. The perpendicular vibrations in addition show a Q branch, for which $\Delta J = 0$. For symmetric top molecules, both parallel and perpendicular vibrations give rise to a Q branch. In molecules of even less symmetry, the spectra become more complex.

6.12 Low-resolution infrared spectra

The infrared spectra of small molecules in the gaseous state provide a great deal of quantitative structural information, but the spectra of large complex molecules in the solid or liquid state, for which the rotational and vibrational structure cannot be fully resolved, have proved equally useful for chemists. The normal modes, which in a large molecule can be very numerous, can be divided into those that involve the whole molecule, termed *skeletal* vibrations. Skeletal vibrations involve many atoms and are associated with a large effective mass and occur at low frequency. They are often referred to as *fingerprint bands*, as they are usually characteristic of the molecule as a whole and can be used to identify it. Thus, a chemist synthesising a molecule can quickly confirm its identity from its infrared spectrum. This region of the spectrum usually occurs in the range 1400–700 cm^{-1}. An example of this use of infrared spectroscopy is given in Fig. 6.15. At higher frequencies, we can observe bands, called *group* vibrations, which are characteristic of various molecular groupings (Table 6.4). These can be used as an analytical tool to help determine the chemical composition of an unidentified molecule.

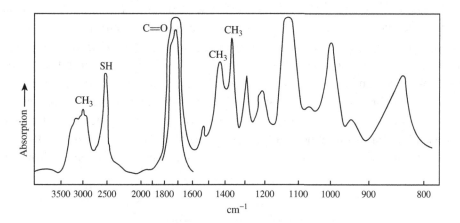

Fig. 6.15. The infrared spectrum of CH$_3$CO.SH.

Table 6.4. Characteristic approximate frequencies in infrared spectra (cm^{-1}).

$\equiv C - H$	3300	$-OH$	3600
$= CH_2$	3030	$-NH_2$	3400
$-CH_3$		$-SH$	2580
Stretch	2950		
Bend	1450		
$>C = O$	1700	$\equiv C - Cl$	700
$>C = C<$	1650	$>C = S$	1100
$-C \equiv N$	2100		

6.13 Raman spectra

When monochromatic radiation, usually generated by a laser, is directed at a substance, the light will be scattered. Most of the scattering occurs without any change of wavelength, but, in 1928, Chandrasekhara Raman, discovered that, under some circumstances, a small fraction was scattered at different wavelengths from the incident beam. The shifts in wavelength were found to be characteristic of the scattering molecules. The lines observed at a lower frequency than the incident radiation are referred to as *Stokes* lines and those at higher frequency as *anti-Stokes* lines. Usually, light in the visible or ultraviolet region is employed and the scattered radiation observed at right angles to the main beam. Raman spectroscopy is an important addition to the chemist's armoury. It follows different selection rules to infrared spectra and the information that can be obtained often complements that obtained from other spectroscopic techniques. This is because transitions become active in Raman spectra when a change in polarisability occurs, unlike infrared, where a change in dipole moment is required. The polarisability of a molecule is the measure of the ease with which an external field can induce a dipole moment in the molecule. The rotational selection rule in Raman spectra is $\Delta J = 0, \pm 2$ (Fig. 6.16) and $\overline{v} = \overline{v}_{inc} \pm B(4J + 6)$, where $\overline{v}_{inc}$ is the wavenumber of the incident radiation. This leads to spectral lines separated by $4B$. The $+$ sign refers to the anti-Stokes lines and the $-$ sign to the Stokes lines. For a molecule to exhibit a pure rotational Raman spectrum, the polarisability must be anisotropic, that is, having different values in different directions. Symmetric molecules, such as CH_4, have polarisability that is the same in all directions and do not have a rotational Raman spectrum. On the other hand, a diatomic molecule, such as nitrogen, will have a

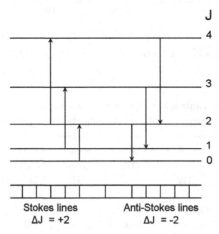

Fig. 6.16. The pure rotational Raman spectrum of a diatomic molecule. A symmetrical pattern of lines is observed each side of the light scattered at the exciting frequency (the Rayleigh scattering).

different polarisability along the inter-nuclear axis than that at right angles to it and the polarisability along the inter-nuclear axis will change as it vibrates. Thus, the vibrational transition of a homonuclear diatomic molecule will be active in Raman spectra but, because there is no change in dipole moment as it vibrates, inactive in the infrared.

6.14 Molecular electronic spectra

The electronic states of diatomic molecules are described using a notation that has much in common with that used for atoms. We define a quantum number, Λ, which determines the projection of the orbital angular momentum about the inter-nuclear axis. $\Lambda = \lambda_1 + \lambda_2 + \cdots$ where for σ orbitals, $\lambda = 0$, and, for π orbitals, $\lambda = 1$ and so on. Thus, Λ is a quantum number in many ways analogous to L in atomic spectra. For $\Lambda = 0, 1, 2, 3, \ldots$, the states are labelled Σ, Π, Δ, Φ, ... The degeneracy of states is again $(2S+1)$, where now S is the component of electron spin angular momentum about the inter-nuclear axis. The resultant total angular momentum about the inter-nuclear axis, analogous to J in atomic spectra, is labeled Ω, where $\Omega = \Lambda + S$. Thus, the designation $^3\Sigma$ would refer to a triplet sigma state, for which $\Lambda = 0$, $S = 1$ and $(2S + 1) = 3$. The states are also labelled according to their symmetry. If the total wave function is symmetrical with respect to inversion through a centre of symmetry of the molecule, it is called a g (gerade) state (Section 5.4). If it is anti-symmetrical, it is labeled u (ungerade). To further specify the symmetry of Σ states, we need to consider whether the wave function changes sign when reflected in a plane through the nuclei. States which change sign are labelled $(-)$ and those which remained unchanged by this reflection are labelled $(+)$. The selection rules for electronic transitions in diatomic molecules are

$$\Delta\Lambda = 0, \pm 1, \quad \Delta S = 0 \quad \text{and} \quad \Delta\Omega = 0, \pm 1.$$

For transitions between Σ states, we have the additional rules $+ \rightarrow +$ and $- \rightarrow -$, and, for homonuclear diatomic molecules, we have g $\rightarrow$ u and u $\rightarrow$ g (Table 6.5).

Table 6.5. Selection rules for diatomic molecules.

Rotation: $\Delta J = \pm 1$
Vibration: $\Delta v = \pm 1, \pm 2, \pm 3, \ldots$ but $\Delta v = \pm 1$ most intense
Electronic: $\Lambda = \Sigma l_i$, $S = \Sigma s_i$ where l_i, and s_i are the components of angular momentum along the internuclear axis and $\Omega = \Lambda + S$.
$\Delta\Lambda = 0, \pm 1$, $\Delta S = 0$ and $\Delta\Omega = 0, \pm 1$
$+ \rightarrow +$ $- \rightarrow -$
$g \rightarrow u$ $u \rightarrow g$

As in atomic spectra, the coupling between the various sources of molecular angular momentum may vary according to the mass of the molecules involved and other factors. The quantum numbers that are used to define the electronic states must be good quantum numbers; i.e. quantum numbers which define properties that are quantised. There are various possible ways in which the angular momenta can combine. The rules governing these combinations are known as Hund's coupling rules and are described in more advanced treatments.

The electronic spectra of diatomic molecules exhibit many of the features we have observed in the infrared region. Electronic transitions alone would appear as a line; however, the accompanying vibrational transitions give rise to a series of bands and, under appropriate circumstances, these bands can be resolved to reveal the rotational structure. The set of bands arising from an electronic transition is referred to as the band system of the transition. Due to the large separation of energy of the vibrational and electronic energy levels compared to thermal energies, almost all absorption spectra arise from molecules in the ground electronic and vibrational state. The intensities of the bands are determined by the *Franck–Condon principle*, based on the Born–Oppenheimer approximation, which states the nuclei move much more slowly than electrons and, as electronic transitions occur much more rapidly than vibrational motion, the transitions can be represented by vertical lines on potential energy diagrams (Fig. 6.17). The intensity is then determined by the Franck–Condon factor, which depends on the values of the vibrational wave functions in the upper and lower states at the same inter-nuclear distance (where one lies directly above the other on the energy versus inter-nuclear distance plot). The factor is defined by the square modulus of the overlap integral, $| \int \psi_1 \psi_2 d\tau |^2$. Approximately, we can say that the maximum value of the wave functions will be near the centre of the ground state and near the turning points of excited states. We can see from the diagram that a change of inter-nuclear distance between the lower and upper states will produce distinctly different patterns of intensity in the resulting bands. In some cases, excitation will be to an unstable upper state which will dissociate. In this case, the spectrum will be a continuum. Sometimes, excitation can occur to an upper state which can then cross over to an unstable excited state. If the change takes place more rapidly than rotation ($\sim 10^{-11}$ s), but less rapidly than a vibration ($\sim 10^{-14}$ s), a well-defined band will be observed, but with no rotational fine structure. This process is called *pre-dissociation*.

We have noted earlier that vibrational energy levels can exhibit anharmonicity. This results in only modest corrections to the predictions obtained by assuming harmonic oscillation in the infrared spectra where only the lowest vibrational levels can be observed. However, in the band structure in electronic spectra, higher

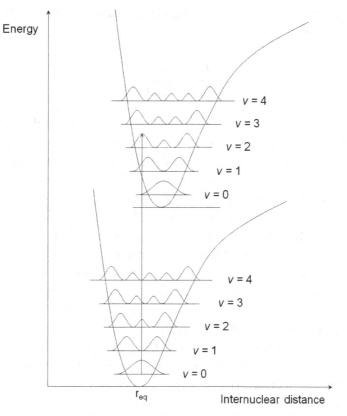

Fig. 6.17. Electronic transitions from the ground state of a diatomic molecule to an excited state. The transitions are represented by a vertical line indicating that no change of separation occurs during the transition, as the electronic motion is much more rapid than the change of inter-nuclear distance caused by vibration (the Born–Oppenheimer approximation). The strongest transitions occur between states where the wave function (and the associated probability of finding the molecule) is greatest. This is the Franck–Condon principle.

vibrational states can be observed and anharmonicity becomes important. The expression for vibrational levels with a single anharmonic term is

$$\frac{E_v}{hc} = \left(v + \frac{1}{2}\right)\omega_e - \left(v + \frac{1}{2}\right)^2 \omega_e x_e.$$

This suggests that the separation between adjacent levels will decrease linearly. Eventually, at dissociation, the separation will become zero and

$$\frac{d\left(\Delta E_v/hc\right)}{dv} = \omega_e - 2\omega_e x_e\left(v + \frac{1}{2}\right) = 0, \quad \text{giving } \left(v + \frac{1}{2}\right)_{dis} = \frac{\omega_e}{2\omega_e x_e}.$$

Substituting this value for $(v+1/2)_{dis}$ in the equation for the vibrational energy gives the dissociation energy as

$$\frac{D_e}{hc} = \frac{\omega_e^2}{4\omega_e x_e}.$$

This method of determining dissociation energies is known as the *Birge–Sponer extrapolation*. As only the leading term of the anharmonicity correction is included, the extrapolation usually gives an overestimate. The dissociation energy can be obtained more accurately by evaluating the area under the curve ΔE_v vs. v (Fig. 6.18) which makes allowance for higher contributions to anharmonicity.

The rotational structure of the bands in electronic spectra differs from that observed in the infrared region because the significant difference in inter-nuclear separation which can occur between the two electronic states gives rise to considerable distortion of the simple rotational pattern. An additional term, which depends on the difference in inter-nuclear separation, occurs in the expression for the rotational energy and can lead to the rotational lines becoming increasingly close together and eventually turning back on themselves, producing what is called a band head. If the inter-nuclear distance in the upper state is greater than that in the lower state, that is, $r_2 > r_1$, then the additional term is found to be negative and the rotational lines in the R branch move closer together and can eventually turn back on themselves forming a *band head* which lies on the high-frequency side of the band origin, $\bar{v}_0$ (Fig. 6.19). The 'concentration' of rotational lines will be less towards the low-frequency end of the spectrum and the band will be said to be *shaded*, or *degraded*, to the red (lower frequencies). If $r_2 < r_1$, the band head will occur in the

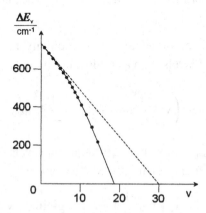

Fig. 6.18. The Birge–Sponer extrapolation for an excited state of oxygen. The area under the curve is the dissociation energy. A linear extrapolation (which is equivalent to the equation above) leads to a significant overestimate, as it ignores higher order anharmonicity terms.

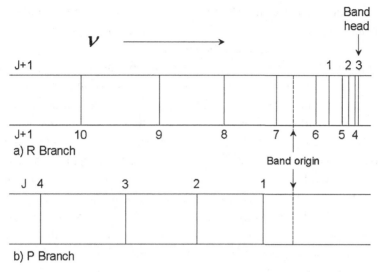

Fig. 6.19. The formation of a band head which arises from a significant difference in the inter-nuclear distance between the upper and lower electronic states. In the case illustrated schematically, the bond length in the upper state is greatest. The separation of the rotational lines in the R branch decreases until, instead of the wavenumber of the rotational lines increasing with increasing J, it begins to decrease. As the density of rotational lines decreases towards the low-frequency end of the spectrum, the band is said to be *degraded(or shaded) to the red*. In contrast, the separation of the lines in the P branch increases. The two branches overlap each other. J is the rotational quantum number for the lower electronic state.

P branch and the band will be shaded towards the violet (high-frequency) end of the spectrum. This means that one can tell by inspection of a spectrum whether the upper or the lower state has a larger inter-nuclear separation.

6.15   Low-resolution electronic spectra

The detailed analysis that can be performed on the electronic spectra of diatomic and other simple molecules is not possible for large polyatomic molecules. Nevertheless, a study of the electronic spectra can still give important information.

Transitions in organic molecules (other than those that have conjugated bonds) usually result from the promotion of an electron from a σ, π, or a non-bonding, n, orbital in the ground state to a vacant anti-bonding σ^* or π^* orbital. The most common transition is a promotion of an electron from a bonding π molecular orbital to an anti-bonding π^* orbital, written $\pi \rightarrow \pi^*$. In the case of unsaturated hydrocarbons, they tend to be the only transitions in the visible and ultraviolet regions. Saturated hydrocarbons lacking π electrons can only exhibit $\sigma \rightarrow \sigma^*$ transitions, which occur in the far ultraviolet regions. Organic molecules may contain non-bonding orbitals, which are generally associated with a heteroatom (an atom other than carbon or hydrogen). The lowest unfilled orbitals are π^* and transitions are designated $n \to \pi^*$. The resulting (n, π^*) states are often the lowest singlet and triplet states of molecules containing heteroatoms.

In conjugated molecules, electrons are highly mobile and the electronic transitions responsible for light absorption can involve orbitals that extend over much of the molecule. Transitions of this sort are important, as they occur in the visible region and the compounds in which they occur include many useful dyestuffs. To estimate the frequency of the transitions, we can make the simple assumption that the electrons are free to range over the length of the conjugated chain. We can model this situation by regarding the electron trapped in a one-dimensional box (see the example of butadiene given in Section 5.7). Calculations of this type, on what is known as the free electron model, explain how conjugated molecules can give rise to spectral transitions at longer wavelengths, in the visible region. However, the equation only gives very broad agreement with the observed spectra and requires empirical adjustments before it can be used to make accurate predictions.

Information about electronic energy levels can also be obtained by *photoelectron spectroscopy*. Molecules are exposed to high-energy radiation and the energy of the electrons ejected analysed. The peaks correspond to the energy of the filled molecular orbitals (Fig. 6.20). The spectra can provide a good test of molecular orbital theory. Photoelectron spectra are interpreted using *Koopmans' theorem*, an approximation which states that the energy of the ejected electrons, less than that of the incident radiation, is equal to the energy of the orbital from which they arise. This approximation takes no account of the fact that removing one electron may cause the others to redistribute and change the energy of the molecule.

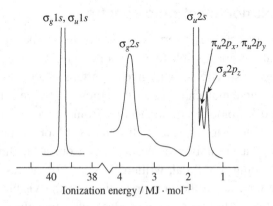

Fig. 6.20. The photoelectron spectrum of nitrogen. The peaks are associated with electrons being ejected from specific molecular orbitals. (See also Table 5.3.) The σ_u and the π_g orbitals are anti-bonding orbitals.

6.16   The fate of excited electronic states

When molecules are raised to an excited electronic state by the absorption of radiation (or by other methods), the excess energy can be released in various ways and the molecule will usually return to its ground state. The mechanisms by which energy is redistributed from the excited state can be either radiative or non-radiative. These processes are illustrated in Fig. 6.21 in what is known as a *Jablonski diagram*.

(i)   *Vibrational relaxation*: Excess energy can be removed non-radiatively by collisions with other molecules. Molecules in higher vibrational levels of an excited electronic state are usually rapidly returned to the lowest vibrational level of that state following collisions.

(ii)   *Internal conversion*: The excited state can convert to a vibrationally-excited ground electronic state.

(iii)   *Intersystem crossing*: A molecule in the first excited singlet state can convert to a lower-lying triplet state.

(iv)   *Fluorescence*: The excited singlet state can fluoresce, with a lifetime on the order of 10^{-8} seconds, losing energy by radiation and returning to the ground state. This process satisfies the selection rule $\Delta S = 0$ so that fluorescence is generally intense and short-lived.

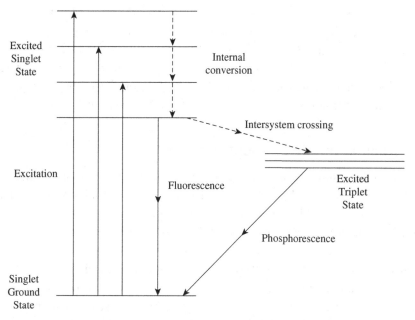

Fig. 6.21.   The Jablonski diagram (schematic), which illustrates the various possible fates of an excited electronic state. The broken lines indicate non-radiative processes.

(v) *Phosphorescence*: Following intersystem crossing, the triplet state can radiate by phosphorescence, returning to the ground state by a transition that is formally forbidden as $\Delta S \neq 0$. This is a slow process leading to weak radiation with a long lifetime, typically in the range 10^{-3} s to many seconds. Phosphorescence occurs at longer wavelengths than fluorescence.

(vi) *Slow fluorescence*: Molecules in the triplet state can be thermally excited back to the upper singlet state and from there decay to the ground state. This process of slow fluorescence has the same spectral characteristics as normal fluorescence, but is usually very sensitive to temperature.

(vii) *Chemical reaction*: The excited state may undergo chemical changes by dissociating or taking part in a chemical reaction. This is the subject of photochemistry.

The lowest excited singlet and triplet states of unsaturated molecules play an important part in photochemistry, and the contrasting behaviour of aromatic hydrocarbons and carbonyl compounds is often used to illustrate the electronic factors that determine the relative importance of the various methods by which an excited state can lose energy. Aromatic hydrocarbons are excited by the absorption of light at $\sim$180 nm. This excites a bonding π electron to an anti-bonding π^* level, leading to a (π, π^*) state. Carbonyl compounds absorb at $\sim$250–360 nm when a non-bonding electron on the oxygen atom is excited to a π^* level (though this transition is forbidden by the selection rules), leading to an (n, π^*) state. In general, (n, π^*) states have longer lifetimes, and, in carbonyl compounds, intersystem crossing to the triplet state occurs relatively efficiently. Thus, they phosphoresce from the lowest triplet state and exhibit little or no fluorescence. By contrast, the (π, π^*) states in aromatic hydrocarbons are short-lived and much of the energy of the excited singlet state is released as fluorescence (with a half-life on the order of 10^{-8} s).

6.17 Nuclear magnetic resonance (NMR)

Many nuclei have spin angular momentum and, as they are charged particles, the spin produces a magnetic dipole moment. These magnetic dipoles may orientate in an applied magnetic field to give states of different energy. A nucleus with a spin quantum number I may take on $(2I + 1)$ different orientations with respect to the applied magnetic field. The most common nucleus employed in nuclear magnetic resonance is the proton and we will confine our attention to nuclei like hydrogen with a spin quantum number $1/2$. The nucleus can adopt orientations of $+1/2$ or $-1/2$, indicating it is aligned with or against the applied field. (A stricter interpretation is that the spins rotate about the field direction, in a process called *Larmor precession*, with frequency corresponding to the separation in energy levels.) The difference in energy between the two orientations in a magnetic field of strength B is

$$\Delta E = g\mu_N B,$$

where g is a factor characteristic of each nucleus, which for protons equals 5.586. μ_N is the *nuclear magneton* which is defined in terms of the mass, m_p, and charge of the proton:

$$\mu_N = \frac{eh}{4\pi m_p} = 5.051 \times 10^{-27} \, \mathrm{J\,T^{-1}}.$$

The energy absorbed by the transition accompanying the 'spin flip', the reversal of direction of the nuclear spins, is given by $\nu = g\mu_N B/h$ and is proportional to the field strength (Fig. 6.22).

The simplest nuclear magnetic spectrometer is illustrated schematically in Fig. 6.23. A sample is placed between the poles of a strong electromagnet which produces a magnetic field that can be adjusted by varying the current through the coils of the magnet. The sample is irradiated by radio-frequency radiation and the amount

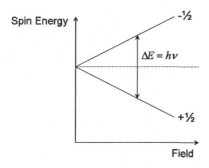

Fig. 6.22. The energy of a proton spin in an applied magnetic field. The splitting is directly proportional to the magnitude of the applied field.

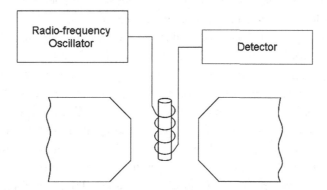

Fig. 6.23. Schematic diagram of a nuclear magnetic resonance spectrometer.

of radiation absorbed by the sample is measured using a radio-frequency detector. A frequency of 60 MHz was traditionally employed and the proton resonances (when the absorption of the radiation was observed) occurred at a magnetic field strength of approximately 1.4 T. The energy difference between the two proton alignments in a field of this magnitude is on the order of 10^{-26} J — very small compared with the thermal energy (kT), which is on the order of 10^{-21} J. Applying the Boltzmann factor, we find that the upper state population will differ from the lower state population by only one part in 10^5. The applied radiation will then stimulate almost equal upward transitions between the two states as downward transitions leading to a very low value of net absorption. Modern spectrometers use superconducting magnets to generate much higher fields of over 10 T, with correspondingly larger energy differences between the two spin states.

The importance of nuclear magnetic resonance to chemists is due to the fact that each nucleus experiences a field slightly different from the applied magnetic field. Small fields are generated by the motion of neighbouring electrons in the applied field. These fields are usually in the opposite direction to the applied field (but this is not always so) and typically on the order of 10^{-5} of the applied field. This shielding means that the frequency at which radiation is absorbed by the nucleus, the *resonance frequency*, is different depending on the molecular environment of the nucleus. This shift in frequency is referred to as the *chemical shift*, δ. In order to measure this frequency shift in a convenient way, it is usual to refer it to a standard, the resonance frequency for the protons in tetramethylsilane $Si(CH_3)_4$, written ν_0. The chemical shift is defined

$$\delta = \left[\frac{(\nu - \nu_0)}{\nu_0} \right] \times 10^6 \text{ ppm},$$

which has the distinct advantage of being independent of the applied field. When a nuclear magnetic resonance spectrum is observed, the resonance for the protons in, for example, a methyl group, will occur at a different frequency from that in, for example, a hydroxyl group. Typical chemical shifts are listed in Table 6.6.

A further feature of the spectrum is that individual lines which exhibit the same chemical shift can show a fine structure, because each magnetic nucleus contributes to the field experienced by other nuclei. The strength of this interaction is much less than that which gives rise to the chemical shift and is referred to as the *spin–spin coupling* and measured by the shift of the frequency. Unlike the chemical shift, its magnitude is not dependent on the strength of the applied field. We can illustrate the effects of chemical shifts and spin–spin coupling by examining the spectrum for ethanol, CH_3CH_2OH (Fig. 6.24). The single hydroxyl proton is comparatively distant from the other protons and produces a single line. The three protons in the methyl group, CH_3, are equivalent and equivalent protons do not affect each

Table 6.6. Chemical shifts in proton magnetic resonance. Relative to $(CH_3)_4Si = 0$ and expressed in ppm.

RC$\underline{H}_3$	0.8–1.0
R$_2$C$\underline{H}_2$	1.2–1.4
R$_3$C$\underline{H}$	1.4–1.6
Ar$\underline{H}$	6.0–8.5
C$\underline{H}_3$Cl	3.0
C$\underline{H}_3$F	4.3
C$\underline{H}_3$I	2.2
ROC$\underline{H}_2$R	3.3–3.9

R represents an aliphatic group and Ar an aromatic group.

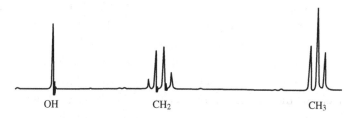

OH   CH$_2$ CH$_3$

Fig. 6.24. The proton nuclear magnetic resonance spectrum of ethanol.

other. (The term equivalent, in this context, is difficult to define rigorously, but simple chemical equivalence will serve our purpose now.) However, these protons will experience the effects of the nuclear spin of the protons in other groups. In particular, the nuclear spins of two protons in the methylene group, $-CH_2$, can be aligned in four possible ways: ↑↓, ↓↑, ↑↑, ↓↓, of which the first two are equivalent. The fields from these spin alignments split the signal from the methyl protons into three lines of which the central one is twice the intensity of the other two. In a similar manner the spin arrangements of the three methyl protons,

$$\uparrow\uparrow\uparrow \quad \uparrow\uparrow\downarrow \quad \uparrow\downarrow\uparrow \quad \downarrow\uparrow\uparrow \quad \uparrow\downarrow\downarrow \quad \downarrow\uparrow\uparrow \quad \downarrow\uparrow\downarrow \quad \uparrow\downarrow\downarrow \quad \downarrow\downarrow\downarrow$$

can be arranged into four groups with relative weighting 1:3:3:1. This leads to the splitting of the line from the methylene group into four. In general, a group of p equivalent electrons splits a neighbouring group as:

$p = 1$	1 1
$p = 2$	1 2 1
$p = 3$	1 3 3 1
$p = 4$	1 4 6 4 1

(This arrangement, where each number is the sum of the two numbers immediately above it, is called Pascal's triangle.) These considerations provide a good explanation of the observed nuclear magnetic resonance spectrum of ethanol if we assume that the hydroxyl group is sufficiently remote not to experience the effects of the methyl and methylene groups.

There is a further factor that complicates NMR spectra. We learned earlier that spontaneous transitions become increasingly rare as energy levels move closer together. In levels as similar in energy as occurs in NMR, spontaneous transitions can be sufficiently rare, and the upper level population can reach that of the lower level so that no net absorption can be observed, a condition known as *saturation*. The observation of an NMR spectrum requires the existence of non-radiative processes that allow molecules to return to the ground state to maintain a difference in the population of the states. This loss of energy from a spin system to the surrounding molecules is called *spin–lattice relaxation* and occurs with a lifetime given the symbol T_1. The uncertainty principle determines that the width of an NMR line is dependent on the lifetime of the spin states; the shorter the lifetime, the broader the NMR line. The lifetime depends on T_1 and also on another relaxation mechanism, in which the spins do not lose energy to their surroundings, but engage in a process which can be thought of, simplistically, as an exchange of their energy with other spins. This process is called *spin–spin relaxation* or *transverse relaxation* and its lifetime is designated T_2. T_2 is typically on the order of a few seconds and, in normal liquids, T_1 is of similar magnitude. Techniques been developed in which pulses of radiation are applied to the sample to enable T_1 and T_2 to be determined independently. In some circumstances, the lifetime can also be affected by chemical exchange processes and the study of line widths in nuclear magnetic resonance provides a convenient method of measuring the speed of certain fast reactions.

Nuclear magnetic resonance is one of the most powerful tools available to chemists. It can be used to identify substances, determine the unknown structure of molecules, measure the rates of chemical processes and, by using non-uniform fields, is employed as an imaging technique of considerable importance in medicine. Nuclear magnetic resonance has been applied to many nuclei other than protons, including ^{13}C and ^{31}P. Fourier transform spectroscopy which, instead of scanning and observing individual frequencies, irradiates the sample with pulses of a wide range of radio frequencies, provides a powerful extension of the method. The absorbed energy will result from all protons in the system absorbing at slightly different frequencies. It is possible by a Fourier transformation to process the responses to construct a normal spectrum. A considerable advantage accrues from being able to look at all frequencies simultaneously.

To delve deeper into the scope of NMR we would need a more thorough approach to the interaction of the nuclear spins with the applied field and, in particular, a more sophisticated representation of the process of spin–spin or transverse relaxation. This is material for a more advanced treatment.

6.18   Electron spin resonance spectroscopy

Electron spins may be studied in a very similar way to nuclear spins. Most molecules do not have unpaired electrons but some molecules (including oxygen and nitric oxide) and free radicals do contain unpaired electrons that can be studied by the technique of electron spin resonance. The electron spins can align with or against the field, in a similar manner to nuclear spins, and we may express the energy difference as a function of the magnetic field strength, B, as

$$\Delta E = g\mu_e B,$$

where g for an electron is a number very close to 2.00 and the Bohr magneton is $\mu_e = 9.274 \times 10^{-24}\,\text{J}\,\text{T}^{-1}$, giving

$$\nu = 2.8 \times 10^{10} B\,\text{s}^{-1}\,\text{T}^{-1}.$$

For fields of 0.3 T the absorption frequencies lie in the region of $10^{10}\,\text{s}^{-1}$ corresponding to an energy difference of $\sim 10^{-23}\,\text{J}$. This energy difference between the electron-spin states is considerably bigger than that between nuclear-spin states and, consequently, the difference in the population between the upper and lower levels is larger. This means that the technique is more sensitive and capable of better resolution. The spectrum is usually observed by irradiating the sample with microwaves of a fixed frequency and adjusting the field to obtain resonance absorption. Electron spin resonance has been particularly useful to chemists for the study of free radicals with an unpaired electron, which play an important part in many chemical processes. In a similar manner to the splitting observed in NMR, the electron spin resonance spectrum can show splittings called hyperfine structure due to the interaction of the unpaired electron with the magnetic fields of nearby nuclei. This can give important information about molecular structures.

6.19   Summary of key principles

Much of this chapter is concerned with the practical application of quantum mechanical principles. The central principle is that the frequency of the electromagnetic radiation absorbed by a transition within a molecule is related to the energy change accompanying the transition by

$$\Delta E = h\nu.$$

The second important principle is that most spectra, other than Raman and magnetic resonance spectra, arise from transitions that are accompanied by a *change in dipole moment* of the substance under investigation. We can often, from symmetry considerations, identify transitions in which the change of dipole moment is nonzero. These spectrally-active transitions can be defined in terms of the corresponding changes in the quantum numbers, which are called the *selection rules*.

Problems

(1) The intervals between a series of lines in the emission spectrum of lithium are observed at 5340, 2472, and 1342 cm^{-1}. Assuming that the spectrum is hydrogen-like, estimate the Rydberg constant and the series limit.

(2) Calculate the energy of the transition from the $J = 0$ to the $J = 1$ rotational state for the hydrogen molecule. The bond length of the hydrogen molecule is 74 pm.

(3) The first three vibrational bands in an excited state of the oxygen molecule have origins at 49363, 50046, and 50710 cm^{-1}. Estimate the dissociation energy of this excited state from the lowest vibrational level. (The accurate value, 7194 cm^{-1}, is significantly different from the value obtained from this calculation. Why is this?)

(4) The separation of lines in the far infrared spectrum of hydrogen bromide is 16.4 cm^{-1}. Calculate the bond length of the molecule.

(5) In the vibrational spectrum of carbon monoxide, a band is observed at 2144 cm^{-1}. Calculate the frequency of the vibration of the molecule and its zero-point energy.

7

The Second Principle: The Higher, the Fewer

The first principle — that of the quantisation of energy — helps us to explain many chemical phenomena. We learn how the number and distribution of electrons in atoms can determine the properties of the chemical elements and the manner in which the elements can form compounds. It also explains the way that light is absorbed by atoms and molecules. In general, it provides us with a considerable understanding of the properties of *isolated* atoms and molecules.

However, we also need to understand how atoms and molecules behave in bulk. When we allow substances to react and observe the result, we are following the behaviour of countless numbers of molecules. This fact, that large numbers of molecules are involved, leads to new challenges. We need to explore the rules that determine where chemical systems will come to a position of equilibrium such that no further change can occur. We will find that the rules which constrain chemical systems appear different from those that determine behaviour in the world of everyday objects. It is this difference in behaviour which we must elucidate.

7.1 Equilibrium

It is helpful to first consider how the position of equilibrium can be understood in mechanical systems. Let us consider a very simple such system: a block of material sliding down a slope, as illustrated in Fig. 7.1 In normal circumstances, it will end up at the lowest point where its potential energy, $U = mgh$, is a minimum. This spontaneous change *could* be used to do work. (We define work as that energy completely convertible into the lifting of a weight — see Section 2.4.) Furthermore, any spontaneous movement of the block will reduce its capacity to do further work.

We can reduce our model to the simple system of two weights over a pulley (Fig. 7.2). This arrangement will be at equilibrium when the forces are balanced and $M_1 = M_2$. It will do the *maximum* amount of work (lifting the weight M_2) when a small change occurs with $M_1 = M_2 + dM$, so that M_2 is virtually equal to M_1. This

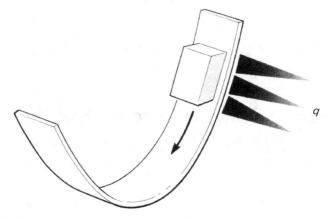

Fig. 7.1. Approach to equilibrium in a mechanical system. Potential energy is dissipated as heat as the block moves to a state of lower energy.

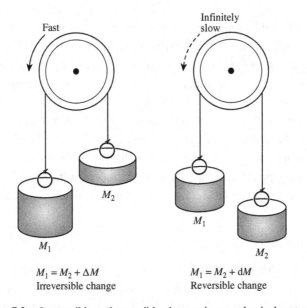

Fig. 7.2. Irreversible and reversible changes in a mechanical system.

process will be, in the absence of finite driving forces, infinitely slow, and is defined as a *reversible* process. The spontaneous processes we see in nature obviously are those which proceed at a discernible rate and are thus *irreversible*, as in our example when $M_1 = M_2 + \Delta M$. *Spontaneous, and therefore irreversible, processes do less than the maximum possible amount of work.*

We can define the position of equilibrium in a number of ways as:

(i) Where the system shows no tendency to change further. (This must be qualified. A mixture of hydrogen and oxygen will appear quite stable until a spark is applied. Then we find out, rather suddenly, that water is the true equilibrium mixture!)

(ii) When the forces acting on the system are balanced.

(iii) When any small change is reversible and leads the system to do the maximum amount of work. This formulation is particularly helpful when we move on from our study of mechanical systems to gain an understanding of physical chemical systems.

7.2 Why we need a second principle

Understanding equilibrium in mechanical systems is quite straightforward and appears to involve little more than common sense. Unfortunately, this is not the case in the world of physicochemical phenomena.

When the block in the dish illustrated in Fig. 7.1 moves to its equilibrium position, where its potential energy is at a minimum, the excess potential energy may be given out as heat (unless the block is harnessed to lift a weight and do work). If chemical processes behaved in a similar manner, we would expect them to liberate heat as they approached equilibrium, which would cause the system to warm up (again assuming they do no work). If we add some solid sodium hydroxide to water, this is indeed what occurs. The equilibrium position, a solution of sodium hydroxide in water, has lower energy, and energy in the form of heat is liberated (Fig. 7.3). Such observations led scientists in the 19th century to suggest, erroneously, that all spontaneous chemical reactions should be accompanied by the evolution of heat (that is, are *exothermic*). However, this is not always the case. If we add solid sodium nitrate to water, we find that heat is *absorbed* and the system cools down (the process is *endothermic*). This system 'climbs uphill' on the energy scale to reach its position of equilibrium (Fig. 7.3). This tells us that, in this system, the tendency to minimise energy cannot be the sole factor determining the position of equilibrium.

Let us consider two systems whose energy is constant and where energy can play no part in determining the position of equilibrium.

(i) *Expansion of a gas.* Consider a system in which a gas is confined to one of two bulbs connected by a stopcock, the other bulb being evacuated (Fig. 7.4). If the stopcock is opened, the gas will distribute itself uniformly between the two vessels. For a perfect gas (and most real gases are almost perfect under normal conditions), there is no change in energy accompanying this expansion.

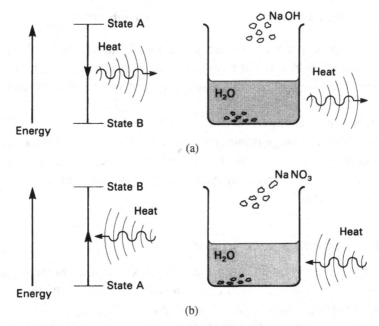

Fig. 7.3. (a) Approach to equilibrium in an exothermic chemical process: heat is evolved. (b) Approach to equilibrium in an endothermic process: heat is absorbed.

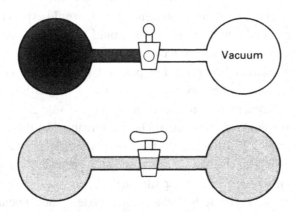

Fig. 7.4. Expansion of a gas into a vacuum.

Nevertheless, there is clearly some driving force causing the gas to distribute itself between the two vessels.

(ii) *Flow of heat.* If we heat a block of hot metal and place it in thermal contact with a colder block, then energy, in the form of heat, will flow until both bodies reach the same temperature. If such a system is insulated from its surroundings, there will be no change in total energy. Again, a property other than energy must determine the approach to equilibrium. The existence of this extra factor explains why, in order to understand equilibrium in physicochemical systems, we require a more sophisticated analysis than sufficed for mechanical systems.

7.3 The second factor

We have observed that a further factor, other than energy, must play a role in determining the approach to equilibrium in physicochemical systems. One example we noted was expansion of a gas from one half of a space to fill the whole. Let us consider this process in terms of the molecules that make up the gas, which we may assume will distribute themselves randomly in the space available to them.

If M gas molecules, first contained in one half of a vessel, as illustrated in Fig. 7.5, are then allowed to occupy the whole volume, the probability that the initial state A could occur by chance is $(1/2)^M$. This is the same as the chance that M objects will all be in one of two boxes between which they have been randomly distributed. We can write the relative probability of state A with respect to state B as

$$\frac{p_A}{p_B} = \left(\frac{1}{2}\right)^M.$$

If, instead of making $V_A/V_B = 1/2$, we select arbitrary values for the volumes of the two compartments, we obtain

$$\frac{p_A}{p_B} = \left(\frac{V_A}{V_B}\right)^M \quad \text{and} \quad \ln\left(\frac{p_A}{p_B}\right) = M \ln\left(\frac{V_A}{V_B}\right).$$

The equilibrium configuration, with the gas molecules spread uniformly throughout the container, is determined by the fact that it is the most probable outcome. When we consider this process from a molecular viewpoint, we realise that it is not *impossible* for all the molecules in a container to be in one half at the same time — only *very improbable*. It is very improbable that all the air molecules in

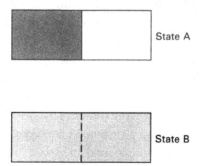

State A

State B

Fig. 7.5. Expansion of a gas. In State A, the gas occupies only half the vessel; in State B, it is uniformly distributed throughout the whole vessel.

the room in which one is sitting will congregate in one corner leaving one with no air to breathe, but it is not, strictly speaking, impossible. To see just how improbable it is, let us consider a mole of gas contained in two volumes, one half the size of the other. Now, for the process B, for 1 mole of gas molecules, $M = N_A$, the Avogadro constant, and

$$\frac{p_A}{p_B} \approx \exp(-10^{23}).$$

This result tells us that it is *extremely unlikely* that the molecules of a mole of gas should all find themselves in one half of the container. (It is even more unlikely that such a disturbing occurrence should occur in one's room, as it will contain many moles of gas.)

7.4 Microstates

To see how we can apply the concept of probability to chemical systems, we must introduce the notion of 'microstates' or 'complexions', defined as the number of ways the overall state of a system can be constructed. The number of ways a state can arise determines the probability of its occurrence. If we toss two coins, the probability of obtaining one head and one tail is twice that of getting two heads. If we write out all the possibilities (HT, TH, HH, TT), we see that this is because two 'microstates' give rise to the mixed configuration.

We can extend these arguments involving probability to a fixed number of molecules distributed over energy levels. The most probable distribution of a large number of objects between a number of boxes, if we have no other constraints, is a uniform distribution, as this would correspond to the maximum number of microstates. However, chemical systems are often constrained; for example, we might have a system in which the total energy is fixed. As an example, let us take three molecules distributed between equally-spaced energy levels (Fig. 7.6) with the total energy fixed at three units. Three distinct distributions, I, II and III, are possible, but not all these distributions are equally probable. We can understand why this is so if we assume the molecules can be individually identified and label each one so as to enumerate the microstates which go to make up each of the three different distributions. We find that six times as many microstates make up II as III, and, thus, it is six times as probable. Configuration I is three times as probable as III.

The number of microstates, W, which make up a distribution is the same as the number of ways N objects can be distributed between a number of boxes so that there are n_0 in the first box, n_1 in the second and so on, which is given by

$$W = \frac{N!}{n_0!n_1!\ldots n_i!}.$$

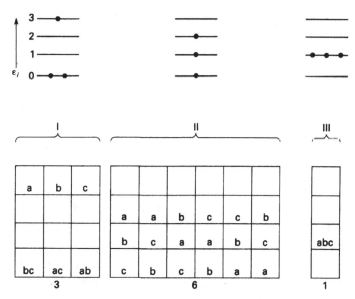

Fig. 7.6. Arrangements of three molecules amongst energy states (with the total energy restricted to three units).

The total number of possible arrangements is $N!$, but we must divide this by $n_0!$, $n_1!$ etc. as we cannot distinguish between states that differ only by an exchange of particles within a given energy level. Thus, for distribution I we have $W_I = 3!/1!2! = 3$, for II, $W_{II} = 3!/1!1!1! = 6$, and for III, $W_{III} = 3!/3! = 1$, as we found by counting all the possible arrangements. This statistical approach enables us to consider how the molecules will be distributed over the energy levels available to them. The distribution they favour at equilibrium is that which tends to maximise W, consistent with the restriction on total energy.

7.5 The Boltzmann factor

When we consider the behaviour of molecules, we can calculate the probability of a particular distribution between the energy levels by calculating the number of microstates, W, as in the previous example. However, there is a significant difference when we are dealing with $\approx 10^{23}$ molecules. To obtain the average of a property over all the possible arrangements, we do not, as would be the case in the previous example, have to consider all distributions. When N is very large, we find that we need only consider the distribution that maximises W and can ignore all others. Thus, to discover the equilibrium configuration, we need only find when

$$W = W_{\text{max}}, \quad \text{where} \quad \left(\frac{\mathrm{d}W}{\mathrm{d}n_i}\right) = 0.$$

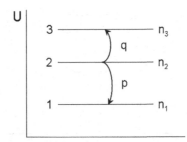

Fig. 7.7. A diagram to illustrate the change in probability when p molecules are moved from level 2 to level 1 and q molecules from level 2 to level 3.

To find the condition under which W is a maximum, let us consider a system of a large number of molecules, N, distributed between a number of energy levels and for which the energy is constant. If we make a small change in the number of molecules in each level, the number of microstates should remain constant when W is a maximum. Turning our attention to just three levels, labelled 1, 2 and 3, we can consider a move of p molecules from level 2 to level 1 and q molecules from level 2 to level 3 (Fig. 7.7) so that the *total energy remains constant*. This requires that p and q and the energies of the three energy levels are related by

$$p(\varepsilon_1 - \varepsilon_2) + q(\varepsilon_3 - \varepsilon_2) = 0$$

and

$$\frac{p}{q} = \frac{(\varepsilon_3 - \varepsilon_2)}{(\varepsilon_2 - \varepsilon_1)}.$$

After the move, the number of molecules in level 2, originally n_2, becomes $(n_2 - p - q)$, and, for level 1, n_1 becomes $(n_1 + p)$ and n_3 becomes $(n_3 + q)$.

If the number of microstates is at a maximum, then these changes must leave W unchanged. Therefore, as

$$W = \frac{N!}{n_1!n_2!\ldots n_i!},$$

for W to be constant, the denominator before and after the molecules are moved must be unchanged, so we obtain

$$n_1!n_2!n_3! = (n_1 + p)!(n_2 - p - q)!(n_3 + q)!$$

Now, as N is, by definition, very large, and p and q are by choice small, $n \approx (n+x)$, and we can use the approximation $(n + x)!/n! \approx n^x$ to obtain

$$\frac{(n_1 + p)!}{n_1!} = n_1^p, \qquad \frac{(n_3 + q)!}{n_3!} = n_3^q$$

and

$$\frac{(n_2 - p - q)!}{n_2!} = n_2^{-[p+q]} = n_2^{-p} n_2^{-q}.$$

Thus,

$$\left(\frac{n_1}{n_2}\right)^p \left(\frac{n_3}{n_2}\right)^q = 1 \quad \text{and} \quad \left(\frac{n_1}{n_2}\right)^p = \left(\frac{n_2}{n_3}\right)^q,$$

giving

$$p \ln\left(\frac{n_1}{n_2}\right) = q \ln\left(\frac{n_2}{n_3}\right).$$

As $p/q = (\varepsilon_3 - \varepsilon_2)/(\varepsilon_2 - \varepsilon_1)$ (see above), we obtain

$$\frac{(\varepsilon_3 - \varepsilon_2) \ln(n_1/n_2)}{(\varepsilon_2 - \varepsilon_1)} = \ln\left(\frac{n_2}{n_3}\right)$$

and

$$\frac{\ln(n_2/n_1)}{(\varepsilon_1 - \varepsilon_2)} = \frac{\ln(n_3/n_2)}{(\varepsilon_2 - \varepsilon_3)}.$$

This equation relates the populations of two energy levels to their difference in energy and we can write it quite generally as

$$\ln\left(\frac{n_2}{n_1}\right) = \beta(\varepsilon_1 - \varepsilon_2) = -\beta \Delta\varepsilon,$$

where β is a constant, obtaining

$$\frac{n_i}{n_0} = \exp(-\beta\varepsilon_i),$$

where ε_i is the energy of the ith energy level measured relative to that of the lowest energy level, ε_0. By comparing the predictions of this equation with the observed properties of molecular systems, it can be shown that

$$\beta = \frac{1}{kT},$$

where k is the Boltzmann constant, $k = 1.381 \times 10^{-23}$ J K^{-1}. The equation governing the occupancy of the energy levels then becomes

$$\frac{n_i}{n_0} = \exp\left(\frac{-\varepsilon_i}{kT}\right).$$

If the energy levels are degenerate, we must allow for this by including a factor g_i to account for the degeneracy of each level and the equation becomes $n_i/n_0 = g_i \exp(-\varepsilon_i/kT)$. This equation was first derived in 1869 by Ludwig Boltzmann. We call $\exp(-\varepsilon_i/kT)$ the *Boltzmann factor* of the ith energy state and the equation the *Boltzmann distribution*. It is one of the most important equations in physical chemistry and provides great insight into the many logarithmic relationships that are so common in chemistry.

At very low temperatures, $\varepsilon_i/kT \ll 1$, and the Boltzmann factors tend to zero. Under these conditions, all molecules will tend to lie in the lowest possible energy state, whereas at very high temperatures the distribution over energy states tends to be more uniform (Fig. 7.8). The energy of a system tends to zero at the absolute zero of temperature and, as all particles then lie in the lowest state, the number of microstates will be unity. We can write the Boltzmann distribution in terms of the probability of an energy state being occupied. If the total number of molecules is N, then the probability of any particular molecule occupying the ith state, p_i, is

$$p_i = \frac{n_i}{N} = \frac{(n_i/n_0)}{(N/n_0)} = \frac{\exp(-\varepsilon_i/kT)}{\Sigma \exp(-\varepsilon_i/kT)},$$

since $N = \Sigma n_i$.

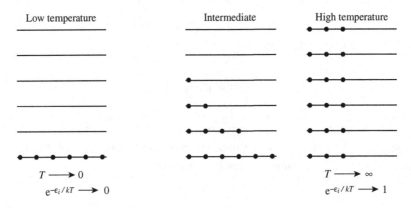

Low temperature   Intermediate   High temperature

$T \longrightarrow 0$

$e^{-\varepsilon_i/kT} \longrightarrow 0$

$T \longrightarrow \infty$

$e^{-\varepsilon_i/kT} \longrightarrow 1$

Fig. 7.8. The distribution of molecules among available energy states at various temperatures.

Although we have obtained the Boltzmann distribution law for systems in which the values of n are large, its validity is more general and it can be shown to apply equally well to systems in which the probability of any particular energy state being occupied is small.

When the number of accessible energy states is less than the number of particles, the Boltzmann distribution law needs to be modified to take account of whether the particles are bosons or fermions. This will tend to occur for particles of very low mass at very low temperatures. For the systems considered in this book, we can almost always consider the distribution law, as given above, to be essentially exact.

7.6 Entropy

The number of microstates, W, is an important factor in determining the equilibrium configuration of a physicochemical system. It is closely related to a thermodynamic quantity, the *entropy* S, that can be related to the number of microstates available to the system using an equation proposed by Boltzmann,

$$S = k \ln W,$$

where k is again the Boltzmann constant. The property entropy, defined in this way, is identical to the thermodynamic function whose importance in determining the position of the equilibrium in molecular systems was recognised as long ago as 1869.

We can investigate how the value of the entropy of the system changes when a small amount of energy is added. Consider a large number of molecules distributed over a series of energy levels with n_i molecules occupying the ith level (Fig. 7.9). If energy ε_i is added to the system to promote one molecule from the lowest level to level i, the number of microstates would change from

$$W_A = \frac{N!}{n_0! n_1! \ldots n_i!} \quad \text{to} \quad W_B = \frac{N!}{(n_0 - 1)! \ldots (n_i + 1)!}$$

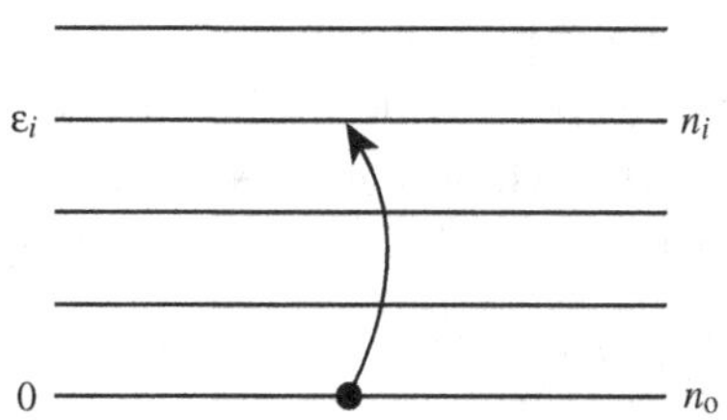

Fig. 7.9. One molecule moved from the lowest to the ith energy state. n_0 and n_i are the numbers of molecules in the two states before the move.

or

$$\frac{W_B}{W_A} = \frac{n_0}{(n_i + 1)}.$$

If $n_i \gg 1$, $W_B/W_A \approx n_0/n_i$.

Since $S = k \ln W$,

$$\Delta S = S_B - S_A = k \ln \left(\frac{W_B}{W_A} \right) = k \ln \left(\frac{n_0}{n_i} \right).$$

However, we have previously established that, for a system at equilibrium, the probability of an energy level being occupied is given by

$$\frac{n_i}{n_0} = \exp \left(\frac{-\varepsilon_i}{kT} \right)$$

and, therefore,

$$\frac{n_0}{n_i} = \exp \left(\frac{+\varepsilon_i}{kT} \right).$$

ε_i is equal to the energy added to the system, which we can write as ΔU. We obtain the change in entropy

$$\Delta S = k \ln[\exp(\Delta U/kT)] = k \left[\frac{\Delta U}{kT} \right] = \frac{\Delta U}{T}.$$

As the energy has been added to the system without any change to the boundaries of the system and, therefore, with no change in the energy levels, we can identify ΔU with the heat added to the system, q (Section 2.4). Thus, we obtain for a small change in the entropy of a system at equilibrium (and for which any small change would be reversible)

$$dS = \frac{dq_{rev}}{T} \quad \text{and} \quad \Delta S = S_B - S_A = \int_A^B \frac{dq_{rev}}{T}.$$

This definition of entropy, though it lacks the insight provided by our definition in terms of microstates, is often more useful, as it does not require any knowledge of the molecular energy levels. It is used to define entropy in chemical thermodynamics — a discipline that explores the interrelation of physicochemical quantities without any reference to molecular structure. This is a subject we will pursue in later chapters.

When $dS = dq/T$, the process is a reversible one and the system must be at equilibrium. This provides us with a very general definition of the position of equilibrium in a system. In an isolated system, which does not exchange energy with its surroundings, $dq = 0$ and the position of equilibrium is defined by $dS = 0$.

We have established that, during a reversible change, a system does the maximum work *on* its surroundings and absorbs the maximum heat *from* its surroundings (Section 7.1). A spontaneous process will do less work than that done in a reversible process, so that $dw > dw_{rev}$. (Remember that we have defined work done *on* the surroundings as negative and so dw is more positive than dw_{rev}.) In an isolated system, $dU = dw + dq = 0$ and, if $dw > dw_{rev}$, then $dq < dq_{rev}$. Thus, for a spontaneous change,

$$dS = \frac{dq_{rev}}{T} > \frac{dq}{T}.$$

For an isolated system, $dq = 0$, and thus $dS > 0$. Spontaneous changes in such a system will result in an increase in entropy until the entropy reaches a maximum value and the system is in equilibrium.

Most systems are not isolated and are free to exchange energy with their surroundings. However, if we consider both the system and the surroundings together, we can regard them as a larger isolated system. Then,

$$dS = dS_{system} + dS_{surroundings} = 0$$

for a reversible process and $dS > 0$ for a spontaneous process. If we extend the definition of system and surroundings to cover the whole universe, then, as spontaneous processes occur, the entropy of this greater system is continually increasing and when it reaches a maximum value no further change will become possible. This has led people to conclude that the energy of the universe is constant, but that the entropy will continually increase until no further change can occur. However, sweeping generalities of this kind lie beyond our current purpose.

In summary, for an isolated system, or a system that is considered together with its surroundings, the position of equilibrium is determined only by the tendency of the entropy to reach a maximum value. This is one of many possible ways of stating the *Second Law of Thermodynamics* (the First Law was introduced in Section 2.5). Though we can express the conditions of equilibrium in this way, it is more convenient to formulate the equilibrium conditions in terms of the properties of the system alone without reference to the surroundings. To do this, we must return to a consideration of systems where the energy may not be constant.

7.7 Defining the position of equilibrium

In the world of mechanical objects, where molecular processes are not involved, the role of the number of microstates, and hence the entropy, is unimportant and the position of equilibrium can be defined solely in terms of minimum energy. Then, the entropy can be regarded as constant and

$$dU = 0 \quad \text{and} \quad U = \text{minimum}.$$

Conversely, for systems at constant energy, the position of equilibrium is determined by the maximum number of microstates and the maximum entropy, so that

$$dS = 0 \quad \text{and} \quad S = \text{maximum}.$$

For most physicochemical systems that we are interested in, the position of equilibrium will be determined by a compromise between these two 'driving forces' — the tendency to minimise energy and the tendency to maximise entropy.

To find the appropriate balance, we must return to the relationships that can be used to characterise the position of equilibrium. The most useful definition is that, at equilibrium, the system has balanced forces and is in a position to do the *maximum work* through a reversible change. We will return to this definition when we give further consideration to equilibrium in chemical systems in the following chapter.

7.8 Entropy as a function of pressure and temperature

Pressure: We have found that when molecules are distributed between two spaces, the number of microstates, W, is directly related to the volume (Section 7.3). Thus, when M molecules of a gas expand from V_A to V_B, we have

$$\frac{W_B}{W_A} = \left[\frac{V_B}{V_A}\right]^M$$

and, as $S = k \ln W$,

$$\Delta S = S_B - S_A = Mk \ln(V_B/V_A).$$

For n moles of gas, $M = nN_A$ and

$$\Delta S = nkN_A \ln\left(\frac{V_B}{V_A}\right) = nR \ln\left(\frac{V_B}{V_A}\right).$$

Since, for a perfect gas, $PV = nRT$, P is proportional to $1/V$ at constant T, and

$$\Delta S = S_B - S_A = -nR \ln\left(\frac{P_B}{P_A}\right).$$

If we define S^0 as the entropy of one mole of perfect gas at one bar pressure (P^0),

$$S = S^0 - R \ln\left(\frac{P}{P^0}\right) = S^0 - R \ln\left(\frac{P}{1\,\text{bar}}\right).$$

Thus, the entropy of a gas decreases as the pressure increases and its volume diminishes.

Temperature: From our definition of heat capacity (Section 2.7), we can write

$$C_V = \left(\frac{\partial q}{\partial T} \right)_V$$

and, as

$$dS = \frac{dq_{\text{rev}}}{T},$$

we obtain

$$dS = \left(\frac{C_V}{T} \right) dT.$$

Integrating this expression from T_A to T_B, assuming C_V is independent of temperature,

$$\Delta S = S_B - S_A = C_V \ln \left(\frac{T_B}{T_A} \right).$$

In a similar manner, we obtain, for the variation of entropy with temperature at constant pressure,

$$\Delta S = C_P \ln \left(\frac{T_B}{T_A} \right),$$

where C_P is the heat capacity at constant pressure.

If S_0 is the entropy at the absolute zero of temperature,

$$S(T) = S_0 + \int_0^T C_P(T) \left(\frac{1}{T} \right) dT = S_0 + \int_0^T C_P(T) d(\ln T).$$

This relationship provides an experimental method of determining the entropy of a substance. The heat capacity is measured as a function of temperature by adding small quantities of heat to the substance, insulated from its environment in a vessel surrounded by a vacuum, and noting the accompanying temperature rise. The apparatus used for these measurements is known as an *adiabatic vacuum calorimeter* (Section 2.9). Since, at absolute zero, atoms or molecules could be expected to lie in their lowest energy level, we can expect that the number of microstates $W = 1$ and that the entropy, S_0, is zero. This is indeed the case for perfect crystals, the equilibrium state of matter at the absolute zero of temperature. The fact that the entropy of substances is zero at the absolute zero of temperature is known as the *Third Law of Thermodynamics*. It enables us to define the *absolute entropy* of substances.

The entropy of chemical substances at room temperature depends very much on their state of matter. It is largely determined by the 'freedom' possessed by the molecules. In solids, when the molecules or atoms are tightly bound, the entropy is low. A rigid solid, such as diamond, has a particularly low entropy ($2.4\,\mathrm{J\,K^{-1}\,mol^{-1}}$). In gases, where the molecules are free to move throughout the whole container, the entropy is high — for CO_2 at one bar and 298 K, its value is $214\,\mathrm{J\,K^{-1}\,mol^{-1}}$. The entropies of liquids lie at intermediate values (Appendix 1). When a substance goes from the liquid state to the gaseous state, in the process of vaporization the entropy will increase as

$$\Delta_{vap}S = \frac{q_{rev}}{T} = \frac{\Delta_{vap}H}{T}.$$

For cyclohexane, $\Delta_{vap}H^0 = 30.1\,\mathrm{kJ\,mol^{-1}}$ and the normal boiling point is 353.7 K. Thus, the entropy of vaporization is $85.0\,\mathrm{J\,K^{-1}\,mol^{-1}}$. For many nonpolar liquids, the entropy of vaporization at their normal boiling point is approximately $85\,\mathrm{J\,K^{-1}\,mol^{-1}}$. This generalisation is known as *Trouton's rule*. The entropy change accompanying chemical reactions depends on the reactants and the products. If solid and liquid reactants produce a gaseous product, the entropy change will be large and positive. With a gaseous reaction in which two molecules combine to form a dimer, the entropy change will be negative.

We should note that entropy, like energy, is a state function and its value depends only on the state of the system. If we go from one state to another the change in entropy is independent of the path we take.

7.9 Partition functions

Having established earlier the factors which determine the probability of an energy level being occupied, we are in a position to calculate the properties of a system in which the energy levels are known. The occupancy of each level is determined by the appropriate Boltzmann factor,

$$\frac{n_i}{n_0} = \exp\left(\frac{-\varepsilon_i}{kT}\right),$$

and the average energy per molecule is

$$\frac{U}{N} = \frac{\Sigma\varepsilon_i n_i}{\Sigma n_i} = \frac{\Sigma\varepsilon_i \exp(-\varepsilon_i/kT)}{\Sigma \exp(-\varepsilon_i/kT)}.$$

It is convenient to identify the denominator as the *molecular partition function*,

$$z = \Sigma \exp\left(\frac{-\varepsilon_i}{kT}\right).$$

The name partition function is somewhat misleading. A better name would be the 'sum over states', but this is not commonly employed. In summing over the energy levels, we have assumed that the levels are not degenerate. This is often not the case and it is more accurate to write the partition function as $z = \Sigma g_i \exp(-\varepsilon_i/kT)$, where g_i is the degeneracy of the ith energy level. However, for simplicity, we will continue to write the partition function in its simple form, but note that the sum must be taken over all *energy states*, rather than all *energy levels*.

The partition function is an extremely valuable function as it can be used to calculate thermodynamic properties. We can see that it is related to the sum of the probabilities of energy levels being occupied and we can regard it as a measure of the *effective number of energy levels* that are accessible to the system at a particular temperature. It will be larger if the energy levels are close together, as is the case of the translational energy levels of gas molecules, and lower if the energy levels are far apart, as are the electronic energy levels in atoms and molecules. The partition function increases as the temperature is raised, as more energy levels are filled.

The expression for the average energy per molecule, given above, can be written in terms of the partition function alone by noting that, at constant volume,

$$\left(\frac{\partial z}{\partial T}\right)_V = \left[\frac{\partial(\Sigma \exp(-\varepsilon_i/kT))}{\partial T}\right]_V = \frac{\Sigma \varepsilon_i \exp(-\varepsilon_i/kT)}{kT^2}$$

and

$$U = \frac{N\Sigma \varepsilon_i \exp(-\varepsilon_i/kT)}{\Sigma \exp(-\varepsilon_i/kT)} = \frac{NkT^2(\partial z/\partial T)_V}{z}.$$

Since $(1/z)\mathrm{d}z = \mathrm{d}(\ln z)$, we obtain

$$U = NkT^2\left(\frac{\partial \ln z}{\partial T}\right)_V.$$

Thus if the energy levels (and their degeneracies) are known, it is possible to calculate U as a function of temperature. Expressing the internal energy, U, in terms of the energy levels and their occupancy enables us to gain further insight into the First Law of Thermodynamics. We have seen that the quantum mechanical solution for the energy of the particle in a box (Section 3.5) shows that the spacing of energy levels depends on the size of the container. We can write

$$dU = \sum \varepsilon_i dn_i + \sum n_i d\varepsilon_i.$$

The change in energy arises from two contributions. The first is a change in the numbers of molecules in each energy level, the energy levels remaining constant with no changes to the boundary of the system. This contribution we can identify as heat. The second term involves changes in the energy levels with the occupation numbers remaining constant. The changes in the energy levels result from changes at the boundary of the system and we identify this contribution as work.

$$dU = dq + dw.$$

Thus, we can understand the concepts of heat and work in molecular terms.

The entropy can be expressed in terms of the partition function starting from the relation (Section 7.6)

$$S = k \ln W,$$

where, for N distinguishable molecules distributed over a set of energy levels,

$$W = \frac{N!}{n_1! n_2! \ldots n_i!} \quad \text{and} \quad \ln W = \ln N! - \Sigma \ln n_i!$$

If N and n_i are large, we can apply Stirling's approximation $\ln N! \approx N \ln N - N$ and obtain $\ln W = N \ln N - N - \Sigma n_i \ln n_i + \Sigma n_i$. Since $\Sigma n_i = N$, we can express this as

$$\ln W = -\Sigma n_i \ln \left(\frac{n_i}{N}\right),$$

giving

$$S = -Nk\Sigma \left(\frac{n_i}{N}\right) \ln \left(\frac{n_i}{N}\right) = -Nk\Sigma p_i \ln p_i,$$

where p_i is the probability that the ith state is occupied.

Since $p_i = \dfrac{\exp(-\varepsilon_i/kT)}{z}$ and $\ln p_i = -\varepsilon_i/kT - \ln z$, we obtain

$$S = -Nk \left[\frac{-\Sigma p_i \varepsilon_i}{kT} - \ln z \Sigma p_i\right].$$

Since $\Sigma p_i = 1$ and $\Sigma p_i \varepsilon_i = U/N$,

$$S = -Nk \left[-\frac{U}{NkT} - \ln z \right] = \frac{U}{T} + Nk \ln z.$$

For one mole, $S = U/T + R \ln z$. Substituting for U in terms of the partition function using the expression obtained above,

$$S = R \ln z + RT \left(\frac{\partial \ln z}{\partial T} \right)_V.$$

In obtaining this expression, we have assumed that the molecules were in some way identifiable. This is true for solids, where the positions of atoms or molecules on lattice sites can be specified. However, it is not true in a gas, where the exchange of two molecules does not produce a distinguishable state. We must, therefore, in the case of perfect gases, divide the number of microstates by $N!$ to allow for the possibility of interchange. This leads to an additional term in the expression for the entropy,

$$-k \ln N! = -kN \ln N + Nk,$$

and

$$S = R \ln z + RT \left(\frac{\partial \ln z}{\partial T} \right)_V - R \ln N_A + R.$$

Other thermodynamic functions can be expressed in terms of the partition function and the formulae are given in Table 7.1.

It is often convenient to express the thermodynamic properties in terms of what is called the *canonical partition function*, Z. This is related to z, the molecular partition function, for an ideal solid by $Z = z^N$ and, for a perfect gas, by $Z = z^N/N!$. We can then write the expressions for internal energy and entropy, $U = kT^2 (\partial \ln Z/\partial T)_V$ and $S = k \ln Z + kT(\partial \ln Z/\partial T)_V$, in terms of the canonical partition function, in a way that applies to both states of matter.

7.10 Determination of thermodynamic functions from partition functions

The total energy of a molecular system can be regarded as the sum of the translational, rotational and vibrational energies, each of which can be calculated if the mass, dimensions and vibrational frequencies of the molecules are known. Since $\varepsilon_i = \varepsilon_{trans} + \varepsilon_{rot} + \varepsilon_{vib}$ and $z = \Sigma \exp(-\varepsilon_i/kT)$, we can write $z = z_{trans} \cdot z_{rot} \cdot z_{vib}$ and can evaluate each of the contributions separately.

Table 7.1. Molar thermodynamic functions in terms of the molecular partition function, z.

Solids	Perfect gases
$U = RT^2 \left(\dfrac{\partial \ln z}{\partial T} \right)_V$	$U = RT^2 \left(\dfrac{\partial \ln z}{\partial T} \right)_V$
$S = R \ln z + RT \left(\dfrac{\partial \ln z}{\partial T} \right)_V$	$S = R \ln z + RT \left(\dfrac{\partial \ln z}{\partial T} \right)_V - R \ln N_A + R$
$A = U - TS = -RT \ln z$	$A = -RT \ln z + RT \ln N_A - RT$
$P = -\left(\dfrac{\partial A}{\partial V} \right)_V = RT \left(\dfrac{\partial \ln z}{\partial V} \right)_T$	$P = RT \left(\dfrac{\partial \ln z}{\partial V} \right)_T$
$H = U + PV$	
$\quad = RT \left[T \left(\dfrac{\partial \ln z}{\partial T} \right)_V + V \left(\dfrac{\partial \ln z}{\partial V} \right)_T \right]$	$H = RT \left[T \left(\dfrac{\partial \ln z}{\partial T} \right)_V + V \left(\dfrac{\partial \ln z}{\partial V} \right)_T \right]$
$G = H - TS$	$G = -RT \left[\ln z - V \left(\dfrac{\partial \ln z}{\partial V} \right)_T - \ln N_A + 1 \right]$
$\quad = -RT \left[\ln z - V \left(\dfrac{\partial \ln z}{\partial V} \right)_T \right]$	$\quad = -RT \ln \left(\dfrac{z}{\cdot N_A} \right)$
$C_V = \left(\dfrac{\partial U}{\partial T} \right)_V = RT \left[2 \left(\dfrac{\partial \ln z}{\partial T} \right)_V + T \left(\dfrac{\partial^2 \ln z}{\partial T^2} \right)_V \right]$	$C_V = RT \left[2 \left(\dfrac{\partial \ln z}{\partial T} \right)_V + T \left(\dfrac{\partial^2 \ln z}{\partial T^2} \right)_V \right]$
$\mu_i = -RT \ln(z_i e^{-Y/RT})$	$\mu_i = -RT \ln \left(\dfrac{z_i}{N_i} \right)$

where Y is the interaction energy of the atoms or molecules in the solid.

The *translational energy levels* are given by (Section 3.5)

$$ E = \left(\frac{h^2}{8m} \right) \left[\frac{n_x^2}{a_x^2} + \frac{n_y^2}{a_y^2} + \frac{n_z^2}{a_z^2} \right]. $$

If we consider only motion in a box of one dimension and length a, we can write

$$ z_x = \sum \exp \left(\frac{-h^2 n^2}{8ma^2 kT} \right). $$

Since translational energy levels are so close together, we can replace the summation by integration to give

$$ z_x = \int \exp \left(\frac{-h^2 n^2}{8ma^2 kT} \right) dn. $$

The integral reduces to the standard form

$$\int_0^\infty \exp(-cx^2)\mathrm{d}x = \left(\frac{\pi}{4c}\right)^{1/2},$$

giving

$$z_x = \left(\frac{2\pi mkT}{h^2}\right)^{1/2} a.$$

Identical contributions result from motion in the other directions and we obtain, for the total translational partition function,

$$z_{trans} = \left(\frac{2\pi mkT}{h^2}\right)^{3/2} a^3 = \left(\frac{2\pi mkT}{h^2}\right)^{3/2} V.$$

We can express this as $z_{trans} = V/\Lambda^3$, where $\Lambda = (h^2/2\pi mkT)^{1/2}$ and has the dimensions of length.

For one mole of gaseous nitrogen at room temperature, we find $z_{trans} = 5.5 \times 10^{30}$, indicating the very large number of translational energy levels available to the molecules.

The *rotational energy levels* of a linear molecule are given by

$$E = \frac{h^2 J(J+1)}{8\pi^2 I}$$

(Section 3.6) and each energy level has a degeneracy of $(2J+1)$. As the rotational energy levels for most molecules are close together compared with kT, we may replace the summation over J by integration, obtaining

$$z_{rot} = \int (2J+1) \exp\left[\frac{-J(J+1)h^2}{8\pi^2 IkT}\right] \mathrm{d}J = \frac{8\pi^2 IkT}{h^2}.$$

This is a correct result for a heteronuclear diatomic molecule, but in order to generalise the result we must add a further factor σ, the *symmetry number*, which accounts for the numbers of ways in which rotation can produce identical configurations (Table 7.2). For a homonuclear molecule, σ is two. The expression for the rotational partition function of a linear molecule becomes

$$z_{rot} = \frac{8\pi^2 IkT}{h^2 \sigma}.$$

Table 7.2. Symmetry numbers.

Molecule	Symmetry	Symmetry number
HCN	linear	1
NO	linear	1
N_2	linear	2
CO_2	linear	2
H_2O	angular	2
NH_3	trigonal pyramid	3
$CH_2=CH_2$	planar	4
CH_4	tetrahedral	12

This can also be written in terms of the rotational constant, B, where

$$B = \frac{h^2}{8\pi^2 I},$$

giving

$$z_{rot} = \frac{kT}{\sigma B}.$$

This is sometimes expressed in an even more concise form, $z_{rot} = T/\sigma\theta_{rot}$, where $\theta_{rot} = h^2/8\pi^2 Ik$. $\sigma\theta_{rot}$ has the dimensions of temperature and is the temperature at which kT becomes equal to the separation between rotational levels (Table 7.3). Evaluating the rotational partition function at room temperature, we find its value for nitrogen is approximately 50, whereas that for chlorine is over 400. Hydrogen,

Table 7.3. Rotational and vibrational energy levels of diatomic molecules.

Molecule	$\Delta\varepsilon_{rot}/\text{J mol}^{-1}$	θ_{rot}/K	$\Delta\varepsilon_{vib}/\text{kJ mol}^{-1}$	θ_{vib}/K
H_2	1420	85	51.0	6140
HCl	254	15	35.8	4300
O_2	34.6	2.1	18.8	2260
N_2	48.1	2.9	28.1	3380
Cl_2	5.8	0.35	6.8	810
Br_2	1.9	0.12	4.0	470
I_2	0.90	0.05	2.6	310
CO	46.0	2.7	26.0	3120
NO	41.0	2.5	22.8	2740

$\Delta\varepsilon_{rot} = \varepsilon_{rot}(J=1) - \varepsilon_{rot}(J=0)$, $\theta_{rot} = (h^2/8\pi^2 Ik)$
$\Delta\varepsilon_{vib} = \varepsilon_{vib}(v=1) - \varepsilon_{vib}(v=0)$, $\theta_{vib} = h\nu/k$
$\sigma\theta_{rot}$ and θ_{vib} are the temperatures at which kT becomes equal to $\Delta\varepsilon_{rot}$ and $\Delta\varepsilon_{vib}$ respectively.

with its very low moment of inertia, has rotational energy levels that are unusually far apart and its rotational partition function is only 1.8. The rotational partition function for a polyatomic molecule is given by

$$z_{rot} = \frac{\pi^{1/2}(8\pi^2kT)^{3/2}(I_xI_yI_z)^{1/2}}{h^3\sigma}.$$

This can also be expressed as $z_{rot} = (\pi^{1/2}/\sigma)(T^3/\theta_x\theta_y\theta_z)$, where $\theta_x = h^2/8\pi^2I_xk$ etc.

The energy levels of *a molecule vibrating with simple harmonic motion* are given by $E = (v + 1/2)h\nu$ (Section 3.7) and, to obtain the partition function, we sum from the lowest energy level, $h\nu/2$

$$z_{vib} = \sum \exp\left(\frac{-vh\nu}{kT}\right)$$

or $z_{vib} = 1 + \exp(-h\nu/kT) + \exp(-2h\nu/kT) + \exp(-3h\nu/kT)$, where v is the vibrational quantum number and ν the vibrational frequency. This is the geometric series $1 + x + x^2 + x^3 + \ldots$ for which the sum to infinity for $x < 1$ is $(1 - x)^{-1}$, and we obtain

$$z_{vib} = \left[1 - \exp\left(\frac{-h\nu}{kT}\right)\right]^{-1}$$

or $z_{vib} = [1 - \exp(-\theta_{vib}/T)]^{-1}$, where $\theta_{vib} = h\nu/k$ is the temperature at which kT equals the energy of the first vibrational transition. Usually, $h\nu/kT > 1$ and the vibrational energy levels are far apart compared with kT, so the vibrational partition function is close to unity.

The expressions we have obtained for the contributions to the partition function from translational, rotational and vibrational energies can be used to calculate the energy and entropy of substances. The contribution of translational motion to the energy is, as the translational energy levels are so close together compared with kT, the classical value $(3/2)RT$. For the same reason, the rotational contribution to the molecular energy for molecules, other than linear molecules containing hydrogen, has the classical value for a linear molecule of RT and, more generally, for polyatomic molecules $(3/2)RT$. The contribution of vibrational motion to the internal energy of a diatomic molecule, measured relative to the zero-point energy of the molecule, is usually negligible. Differentiating the expressions for energy with respect to T, we obtain, for the heat capacity C_V of a diatomic molecule, a value of $(5/2)R$. For a perfect gas, $C_P = C_V + R$ (Section 2.7) and $C_P = (7/2)R$. These results are summarised in Table 7.4.

Table 7.4. Summary of the properties of linear molecules.

Mode	Partition function, z	Internal energy, $(U - U_0)$	Entropy, S	Heat capacity, C_V
Translation	$V/\Lambda^3 \sim 10^{30}$	$3/2\,RT$	$R\ln(V/\Lambda^3 N_A) + 5/2\,R$	$3/2\,R$
Rotation	For $T > \theta_{rot}$ $T/\theta_{rot} \sim 10 - 100$	RT	$R\ln(T/\sigma\theta_{rot}) + R$	R
Vibration	$[1 - \exp(-\theta_{vib}/T)]^{-1} \sim 1$	See Table 7.5	See Table 7.5	See Table 7.5
Electronic	g_{elec}		$R\ln g_{elec}$	

$\Lambda = (h^2/2\pi mkT)^{1/2}$, $\theta_{rot} = h^2/8\pi^2 Ik$, $\theta_{vib} = h\nu/k$

The contribution of the various modes to the entropy of a molecule can be evaluated using the relationship given in Section 7.9. Expressed in convenient units, we obtain, for the contributions to the entropy of a mole of substance where M_r is the relative molar mass (molecular weight),

$$\frac{S_{trans}}{R} = \frac{5}{2}\ln\left(\frac{T}{K}\right) + \frac{3}{2}\ln(M_r) - \ln\left(\frac{P}{bar}\right) - 1.152$$

and, for a linear molecule,

$$\frac{S_{rot}}{R} = \ln\left(\frac{I}{amu\,nm^2}\right) + \ln\left(\frac{T}{K}\right) - \ln\sigma + 2.418.$$

The contribution of vibrational motion to entropy is usually very small.

If we examine these expressions in more detail as a function of temperature over a wide range, we see that each mode at appropriately low temperatures makes a contribution of effectively zero to the energy. At high temperatures, the contribution of each mode to the energy approaches the classical value $RT/2$ and the contribution to the heat capacity $R/2$ (Fig. 7.10). This transition can be observed with both rotational and vibrational contributions if measurements at sufficiently high and low temperatures are available. The thermodynamic properties of a harmonic oscillator as a function of temperature are given in Table 7.5.

We can apply the results obtained above to the calculation of the thermodynamic properties of gaseous oxygen at 298 K and 1 bar pressure. We have $T = 298$ K, the molar mass $M = 32$ g mol^{-1} and the bond length of oxygen, d, as 0.121 nm. The moment of inertia, I, is given by

$$I = M\left(\frac{d}{2}\right)^2 = 32\left(\frac{0.121}{2}\right)^2 = 0.1171\ amu\ nm^2.$$

The vibrational frequency of the oxygen molecule expressed in wave numbers is 1580 cm^{-1}, corresponding to a value of $h\nu/kT = 7.6$. Inspection of Table 7.5

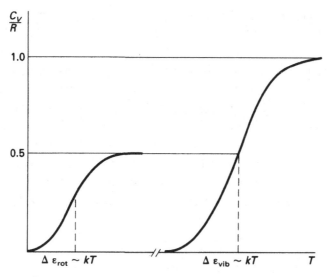

Fig. 7.10. The contribution to heat capacity of a vibrational and rotational degree of freedom. The heat capacity of both modes is zero at the absolute zero of temperature. The vibrational degree of freedom has a maximum heat capacity of R as it possesses both potential energy and kinetic energy.

Table 7.5. Thermodynamic properties of a harmonic oscillator.

$h\nu/kT$	z	$-(G - U_0)/RT$	$(U - U_0)/RT$	S/R	C_V/R
	$(1 - e^{-x})^{-1}$	$\ln(1 - e^{-x})$	$x/(e^x - 1)$		$x^2 e^x/(e^x - 1)^2$
0	∞	∞	0	0	1
0.1	10.508	2.352	0.951	3.303	0.999
0.2	5.517	1.708	0.903	2.611	0.997
0.3	3.858	1.350	0.857	2.208	0.993
0.5	2.541	0.933	0.771	1.703	0.979
0.6	2.216	0.796	0.730	1.526	0.971
0.8	1.816	0.597	0.653	1.249	0.948
1.0	1.582	0.459	0.582	1.041	0.921
1.4	1.327	0.283	0.458	0.741	0.852
1.8	1.198	0.181	0.356	0.537	0.769
2.2	1.125	0.117	0.274	0.392	0.678
2.6	1.080	0.077	0.209	0.286	0.586
3.0	1.052	0.051	0.157	0.208	0.496
4.0	1.019	0.018	0.075	0.093	0.304
5.0	1.007	0.007	0.034	0.041	0.171
10.0	1.000	0.000	0.000	0.000	0.005

$x = h\nu/kT \quad S/R = (U - U_0)/RT - (G - U_0)/RT = x/(e^x - 1) - \ln(1 - e^{-x})$

indicates that the vibrational contribution to both energy and entropy is very small. We obtain $S_{trans} = 152.07\,\mathrm{J\,K^{-1}\,mol^{-1}}$, and $S_{rot} = 43.87\,\mathrm{J\,K^{-1}\,mol^{-1}}$. The oxygen molecule is unusual in having two unpaired electrons leading to a triply degenerate ground electronic state. This leads to an extra contribution to the entropy of $R\ln 3 = 9.13\,\mathrm{J\,K^{-1}\,mol^{-1}}$. The resulting total standard entropy for oxygen at $298\,\mathrm{K}$ and $1\,\mathrm{bar}$ is $205.07\,\mathrm{J\,K^{-1}\,mol^{-1}}$ in excellent agreement with the observed value of $205.14\,\mathrm{J\,K^{-1}\,mol^{-1}}$ (Appendix 1). The internal energy of oxygen has the classical value (for the translational and rotational contributions), $U = 5RT/2 = 6.19\,\mathrm{kJ\,mole^{-1}}$. Calculations of thermodynamic properties from spectroscopic data are an important part of the chemist's armoury. In many circumstances, such calculations can provide more accurate data than experimental thermodynamic measurements.

7.11   Summary of key principles

Whereas, in a mechanical system, the position of equilibrium is determined solely by the energy, in chemical systems a second factor is involved. Such systems can appear, spontaneously, to 'climb uphill' on the energy scale (endothermic reactions) and processes can occur that do work even when the energy is constant. The additional factor that determines the position of equilibrium in chemical systems is related to the number of microstates that are available to the system. This factor is the thermodynamic function, *entropy*, which can be defined as $S = k\ln W$, where W is the number of microstates.

The entropy of a substance increases with temperature and decreases with increasing pressure. Entropy can also be defined in terms of the heat absorbed reversibly by a system, without reference to molecular properties, by

$$\mathrm{d}S = \frac{\mathrm{d}q_{rev}}{T}.$$

In the approach to equilibrium in chemical systems, the internal energy, U, tends to a minimum and the entropy, S, tends to a maximum.

We find that, at equilibrium, the probability of each energy level of a system being occupied is given by

$$n_i = n_0 \exp\left(\frac{-\varepsilon_i}{kT}\right),$$

the *Boltzmann factor*. This is perhaps the most important relationship in physical chemistry. It enables us to explain many of the logarithmic relations that are so characteristic of physical chemistry.

We define *partition functions* by

$$z = \sum \exp\left(\frac{-\varepsilon_i}{kT}\right).$$

If we have a knowledge of the bond lengths and vibrational frequencies of a chemical compound (as can be obtained from molecular spectroscopy), the partition function allows us to calculate thermodynamic properties directly from these molecular properties.

Problems

(1) Calculate the entropy change when 1 dm^3 of a perfect gas is compressed from 1 atm to 100 atm pressure at 298 K.

(2) The heat capacity of liquid water at constant pressure at 298 K is 75.3 J K^{-1} mol^{-1}. Estimate the change of entropy when one mole of water is heated to 360 K.

(3) Calculate the change in entropy when 100 g of benzene ($M = 78$ g mol^{-1}) is vaporized at its normal boiling temperature, 353 K. The enthalpy change on vaporization is 39.4 kJ mol^{-1}. Compare the value with that calculated using Trouton's rule.

(4) Calculate the entropy change when 22 g of ice, at 273 K and 1 bar pressure, is melted to form water at 298 K. The enthalpy change on fusion of ice is 6.02 kJ mol^{-1} and the heat capacity of water is 75 J K^{-1} mol^{-1}.

(5) Calculate the molecular partition function and molar entropy of gaseous krypton at 1 bar pressure and 600 K. ($M = 83.8$ g mol^{-1})

(6) Calculate the rotational partition function of HCl at 298 K and 1 atm pressure. The bond length in hydrogen chloride is 1.27 Å, the molar mass 36.5 g mol^{-1} and the vibrational wave number is 2991 cm^{-1}. (See conversion factor in Appendix 4.)

8
Chemical Equilibrium

The position of equilibrium can be expressed for an isolated system, or for a system and its surroundings, in terms of entropy (Section 7.6). Under these circumstances, the entropy will be a maximum at equilibrium and for any small change $dS = 0$. However, it is usually more convenient to express the criteria for equilibrium in terms of the properties of the system alone without reference to its surroundings, and the systems we deal with in chemistry are often not isolated and may transfer energy to and from their surroundings.

8.1 Free energy

To identify conditions that define the position of equilibrium for systems that are not isolated, we need to return to the concept of maximum possible work. For a reversible change, since $dS = dq_{rev}/T$, we obtain

$$dU = dq_{rev} + dw_{rev} = T\,dS + dw_{rev}.$$

The reversible work is

$$dw_{rev} = dU - TdS.$$

We define a new thermodynamic state function, the *Helmholtz free energy*, A (now often referred to simply as Helmholtz energy), by

$$A = U - TS$$

and, at constant temperature and volume, $dw_{rev} = dU - TdS = dA$.

Since this equality is defined for a reversible process, $dw = dA$ is a condition for the system to be at equilibrium. If, during a reversible change, the system does work, both dw and dA will be negative. The Helmholtz free energy is equivalent, in a system at constant temperature and volume, to the energy in a mechanical system. It is a measure of the maximum amount of work that can be done by the system

175

on its surroundings. For a spontaneous change, the system will do less work and $dw > dA$. Both dw and dA will be negative, but dw will be less negative than dA. Most chemical systems (other than electrochemical cells) do only PV work which, for a system at constant volume, is zero. For such a system, $dw = 0$ and $dA < 0$ and the change in A will again be negative. Eventually, if spontaneous changes occur, A will reach a minimum value and no further work can be obtained from the system. Then, $dw_{rev} = 0$ and $dA = 0$. Thus, the condition for equilibrium for a system at constant temperature and volume which does no work is

$$dA = 0$$

and the Helmholtz free energy will be a minimum.

8.2 Gibbs free energy

Chemists usually perform experiments at constant pressure rather than at constant volume, and it is helpful to define a condition of equilibrium at constant temperature and pressure. Under these conditions, we can write

$$dw = -PdV + dw_{additional},$$

where $dw_{additional}$ is the work, other than PV work, done on the system.

Since $dw_{rev} = dU - TdS$, we obtain, under reversible conditions,

$$dw_{additional} - PdV = dU - TdS.$$

We can define a new state function, the *Gibbs free energy* (also widely referred to as Gibbs energy),

$$G = U + PV - TS = H - TS$$

and

$$dG = dU + PdV + VdP - TdS - SdT.$$

At constant temperature and pressure, when $dT = 0$ and $dP = 0$, we obtain

$$dG = dU + PdV - TdS$$

and, since

$$dw_{additional} = dU + PdV - TdS,$$

we have

$$dw_{\text{additional}} = dG.$$

This is a condition for equilibrium in a system of constant temperature and pressure.

For systems at constant temperature and pressure doing no additional work, the most common circumstance in chemistry, at equilibrium we have

$$dG = 0$$

and the Gibbs free energy, G, will be at a minimum.

In the equation $G = H - TS$, we have found the proper balance between the tendency to minimise energy (or, at constant pressure, more strictly the enthalpy) and the tendency to maximise entropy when approaching equilibrium. At low temperatures, the tendency to minimise energy is the most important factor, whereas at higher temperatures the tendency to maximise entropy is dominant.

For example, consider the equilibrium

$$\text{Cl}_2(\text{g}) \rightleftharpoons 2\,\text{Cl}(\text{g}).$$

The standard enthalpy change, $\Delta_r H^0$, for the reaction is positive. (When we give the changes in the thermodynamic functions accompanying an equilibrium reaction such as the above, they are for the reaction proceeding to completion, from left to right as written, with the reactants and products under standard conditions.) That is, $\Delta_r H^0$ is negative in the direction of the formation of Cl_2 in which the atoms are held together by a strong binding energy. On the other hand, the two gaseous chlorine atoms possess more entropy than the diatomic chlorine molecule. Thus, as the temperature increases, the tendency of Cl_2 to dissociate becomes greater.

We have now established how the position of equilibrium is determined at constant temperature and pressure. This is a result of considerable practical importance to chemists. In fact, knowledge of how the Gibbs free energy varies with temperature and pressure can tell us how the position of a chemical equilibrium will depend on these variables.

8.3 The pressure and temperature dependence of Gibbs free energy

From the definition of the Gibbs free energy, G

$$G = U + PV - TS$$
$$dG = dU + PdV + VdP - TdS - SdT$$

but, for a system that does only PV work,

$$dU = dq_{rev} + dw_{rev} = TdS - PdV,$$

thus

$$dG = VdP - SdT.$$

This is a most important equation for chemists, as we can now illustrate.

At constant temperature,

$$dT = 0 \quad \text{and} \quad dG = VdP \quad \text{or} \quad \left(\frac{\partial G}{\partial P}\right)_T = V.$$

For n moles of a perfect gas, $PV = nRT$, so that

$$dG = nRT\left(\frac{dP}{P}\right).$$

For a change of pressure from P_A to P_B, we can integrate this equation to obtain

$$\Delta G = nRT\ln\left(\frac{P_B}{P_A}\right).$$

The free energy of a gas is usually expressed in terms of the standard Gibbs free energy, G^0, which is defined as a free energy of one mole of gas at one bar pressure (and most frequently at 298 K). (Remember that one bar is, for most practical purposes, the same as one standard atmosphere.)

Then,

$$G = G^0 + RT\ln\left(\frac{P}{P^0}\right) = G^0 + RT\ln\left(\frac{P}{1\,\text{bar}}\right).$$

Returning to our basic equation $dG = VdP - SdT$ at constant pressure, $dP = 0$ and

$$dG = -SdT \quad \text{and} \quad \left(\frac{\partial G}{\partial T}\right)_P = -S.$$

Since

$$G = H - TS, \quad G = H + T\left(\frac{\partial G}{\partial T}\right)_P.$$

Rearranging and dividing by T^2,

$$\frac{-G}{T^2} + \frac{1}{T}\left(\frac{\partial G}{\partial T}\right)_P = -\frac{H}{T^2}.$$

The left-hand side of the equation is equal to $[\partial(G/T)/\partial T]_P$, giving

$$\left[\frac{\partial(G/T)}{\partial T}\right]_P = \frac{-H}{T^2} \quad \text{and} \quad \left[\frac{\partial(\Delta G/T)}{\partial T}\right]_P = \frac{-\Delta H}{T^2}.$$

These equations are known as the *Gibbs–Helmholtz equations*. They are important to chemists because they relate the temperature dependence of free energy, and hence the position of equilibrium, to the enthalpy change accompanying a reaction (see Section 8.5).

8.4 Chemical potential

The free energy of chemical systems will change if we alter the temperature or pressure. However, importantly for chemists, it can also be changed by adding more substances or by changing the composition, as occurs when chemical reactions take place. If we add dn_i moles of a substance i to a system, we can write

$$dG = \left(\frac{\partial G}{\partial n_i}\right)_{T,P,n_j} dn_i,$$

where n_j indicates that the quantities of all substances, other than substance i, are constant. We define the *chemical potential* of substance i, μ_i, by

$$\mu_i = \left(\frac{\partial G}{\partial n_i}\right)_{T,P,n_j}$$

and

$$dG = VdP - SdT + \Sigma\mu_i dn_i.$$

This equation is sometimes called the *fundamental equation of chemical thermo-dynamics*. The chemical potential may be thought of as the change in free energy when one mole of substance i is added to such a large quantity of the mixture that it does not significantly change the composition. It can be regarded as providing a 'force' that drives chemical systems to equilibrium.

Consider a substance i distributed between the two phases, α and β. If the chemical potential in phase α is $\mu_i(\alpha)$ and that in phase β is $\mu_i(\beta)$ then, if we transfer dn_i of a substance i from phase α to phase β,

$$dG = [\mu_i(\beta) - \mu_i(\alpha)]dn_i.$$

At equilibrium, $dG = 0$ and, as we can always consider the transfer of a small but nonzero value of dn_i, this requires that

$$\mu_i(\alpha) = \mu_i(\beta).$$

This is the condition of equilibrium. Thus, *for a system at constant temperature and constant pressure, the chemical potential of every component must be equal in all parts of the system.* For a pure substance, $(\partial G/\partial n_i)_{T,P}$ is simply the molar free energy, G/n_i, since the free energy will increase linearly as substance is added. For one mole of perfect gas, $n = 1$, and

$$G = G^0 + RT \ln \left(\frac{P}{\text{bar}} \right) \quad \text{(Section 8.3)} \quad \text{and} \quad \mu = \mu^0 + RT \ln \left(\frac{P}{\text{bar}} \right).$$

In mixtures of perfect gases, the total pressure is the sum of their partial pressures, the pressures that each of the gases would exert if it were alone in the container. We can write, for each constituent of a perfect gas mixture,

$$\mu_i = \mu_i^0 + RT \ln \left(\frac{P_i}{\text{bar}} \right),$$

where P_i is the partial pressure of component i.

8.5 Equilibrium between gaseous reactants

Consider a very simple example of chemical equilibrium in the gas phase,

$$A(g) \rightleftharpoons B(g).$$

This could represent, for example, the equilibrium between two isomers, such as n-butane and isobutane. If we imagine the reaction to take place so that dn_A moles of substance A are converted into dn_B moles of B at constant temperature and pressure, we have, for the change in free energy,

$$dG = (\mu_A dn_A + \mu_B dn_B),$$

where dn_A is negative and dn_B is positive. We can define an *extent of reaction*, ξ, which is zero when the reaction position is entirely to the left of the equation, that is, where only reactants are present, and is 1 when the reactants have converted over entirely to products.

For this simple example,

$$d\xi = dn_B = -dn_A \quad \text{and} \quad dG = (\mu_B - \mu_A) d\xi \text{ at constant } T \text{ and } P.$$

The reaction will proceed until G reaches a minimum value and $(\partial G/\partial \xi)_{T,P} = 0$, as illustrated in Fig. 8.1. At the equilibrium position, $\mu_A = \mu_B$.

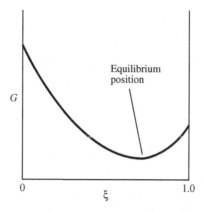

Fig. 8.1. The Gibbs free energy as a function of the extent of a chemical reaction (ξ).

If the two components behave as perfect gases,

$$\mu_i = \mu_i^0 + RT \ln \left(\frac{P_i}{\text{bar}} \right) \quad \text{and}$$

$$\left(\frac{\partial G}{\partial \xi} \right) = \mu_B - \mu_A = \mu_B^0 - \mu_A^0 + RT \ln \left(\frac{P_B}{P_A} \right).$$

$\mu_B^0 - \mu_A^0$ is $\Delta_r G^0$, the free energy change accompanying one mole of reaction (from left to right as written) with both reactants and products remaining in their standard state, that is, at 1 bar pressure and

$$\left(\frac{\partial G}{\partial \xi} \right)_{T,P} = \Delta_r G^0 + RT \ln \left(\frac{P_B}{P_A} \right).$$

$(\partial G / \partial \xi)_{T,P}$ is the change in free energy with the extent of the reaction under the conditions specified by P_A and P_B. It is equal to the free energy change for one mole of reaction with reactant A at a partial pressure P_A going to product B at partial pressure P_B, with the partial pressures remaining constant.

$(\partial G / \partial \xi)_{T,P}$ was traditionally called the *reaction free energy* and written simply as ΔG. We will write it as $\Delta_r G'$ to remember that it is a differential quantity and only corresponds to the Gibbs free energy change for one mole of reaction under precisely defined conditions. In modern treatments, it is called the *affinity* of the reaction.

At equilibrium,

$$\Delta_r G' = \left(\frac{\partial G}{\partial \xi} \right)_{T,P} = 0$$

and, as

$$\Delta_r G' = \Delta_r G^0 + RT \ln \left(\frac{P_B}{P_A} \right),$$

we have

$$\Delta_r G^0 = -RT \ln \left(\frac{P_B}{P_A} \right)_{equil}.$$

The value of (P_B/P_A) at equilibrium is the *equilibrium constant* of the reaction, K_P (Section 1.10), and

$$\Delta_r G^0 = -RT \ln K_P.$$

This important equation tells us how the position of the chemical equilibrium can be defined in terms of the Gibbs free energies of the reactants and products at 1 bar pressure. Such standard free energies can be determined experimentally and are tabulated for use in this way (Appendix 1). If ΔG^0 is positive, the reaction will tend not to proceed and the reactants will predominate in the equilibrium mixture. Conversely, if ΔG^0 is negative, the reaction will proceed and the products will dominate at equilibrium.

Had the reaction been more complicated, the results would be essentially the same. Consider the reaction

$$aA + bB \rightleftharpoons lL + mM,$$

for which

$$dG = \mu_L dn_L + \mu_M dn_M + \mu_A dn_A + \mu_B dn_B + V dP - S dT.$$

The extent of reaction ξ is now defined as

$$d\xi = \frac{dn_L}{l} = \frac{dn_M}{m} = \frac{-dn_A}{a} = \frac{-dn_B}{b} = \frac{dn_i}{v_i},$$

where v_i represents the stoichiometric coefficients $-a, -b, m$ and l. The stoichiometric coefficients for the reactants are negative and those for the products positive.

Thus, we can write, at constant T and P,

$$\partial G = (l\mu_L + m\mu_M - a\mu_A - b\mu_B)\partial\xi$$

and

$$\Delta_r G' = \left(\frac{\partial G}{\partial \xi} \right)_{T,P} = (l\mu_L + m\mu_M - a\mu_A - b\mu_B) = \Sigma v_i \mu_i.$$

Since

$$\mu_i = \mu_i^0 + RT \ln \left(\frac{P_i}{bar} \right),$$

we obtain

$$\Delta_r G' = \Delta_r G^0 + RT \ln \left[\frac{(P_L/\text{bar})^l (P_M/\text{bar})^m}{(P_A/\text{bar})^a (P_B/\text{bar})^b} \right],$$

which we can write more concisely as

$$\Delta_r G' = \Delta_r G^0 + RT \ln \prod \left(\frac{P_i}{\text{bar}} \right)^{\upsilon_i}.$$

At equilibrium, $\Delta_r G' = 0$. Thus,

$$\Delta_r G^0 = -RT \ln K_P,$$

and

$$K_P = \left[\frac{(P_L/\text{bar})^l (P_M/\text{bar})^m}{(P_A/\text{bar})^a (P_B/\text{bar})^b} \right]_{\text{equil}}$$

$$K_P = \left[\prod \left(\frac{P_i}{\text{bar}} \right)^{\upsilon_i} \right]_{\text{equil}}.$$

$\Delta_r G^0 = \Sigma \upsilon_i \mu_i^0$ is the standard Gibbs free energy change for a mole of reaction. We note that K_P is always dimensionless even if $(a + b)$ is not equal to $(l + m)$ as all the pressures comprising it are themselves dimensionless ratios.

8.6 The temperature dependence of the equilibrium constant

Combining the Gibbs–Helmholtz equation (Section 8.3)

$$\left[\frac{\partial (\Delta G/T)}{\partial T} \right]_P = \frac{-\Delta H}{T^2} \quad \text{and} \quad \Delta_r G^0 = -RT \ln K_P,$$

we obtain

$$\left(\frac{\partial \ln K_P}{\partial T} \right)_P = -\frac{1}{R} \left[\frac{\partial (\Delta_r G^0/T)}{\partial T} \right]_P = \frac{\Delta_r H^0}{RT^2}.$$

This important equation is known as the *van 't Hoff isochore* and, since $\Delta_r G^0$ and $\Delta_r H^0$, are, by definition, not functions of pressure (they are the values with the gaseous reactants and products at the standard state of 1 bar), we need not include the restriction to constant pressure, so that

$$\frac{d \ln K_P}{dT} = \frac{\Delta_r H^0}{RT^2}.$$

If we assume that $\Delta_r H^0$ is independent of temperature, which is often a very reasonable approximation, we can integrate the equation to obtain

$$\ln(K_2/K_1) = -\Delta_r H^0[1/T_2 - 1/T_1]/R,$$

where K_1 is the equilibrium constant at temperature T_1 and K_2 is the value at temperature T_2. We can obtain the enthalpy change of a reaction quantitatively from the variation of the equilibrium constant with temperature if we plot $\ln K_P$ as a function of $1/T$ to obtain a straight line of slope $-\Delta_r H^0/R$ (Fig. 8.2).

For an exothermic reaction, which evolves heat, and for which $\Delta_r H^0 < 0$, K_P decreases as the temperature increases. For endothermic reactions, the converse is true and K_P increases with increasing temperature. These observations are in keeping with Le Chatelier's principle (Section 1.10) (see Table 8.1). For the reaction

$$2\,SO_2(g) + O_2(g) \rightleftharpoons 2\,SO_3(g), \quad \Delta_r H^0 = -198\,kJ\,mol^{-1},$$

and we obtain less SO_3 as the temperature is raised. For an endothermic reaction such as

$$2\,HI(g) \rightleftharpoons H_2(g) + I_2(g), \quad \Delta_r H^0 = 53\,kJ\,mol^{-1},$$

and we find more dissociation of HI at high temperatures.

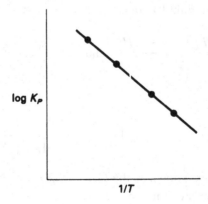

Fig. 8.2. The logarithm of an equilibrium constant as a function of reciprocal temperature.

Table 8.1. Thermodynamic functions and the behaviour of equilibrium constants.

Thermodynamic function	Effect on the equilibrium constant	Effect on the composition of the equilibrium mixture
$\Delta G^0 = +$ve	$K < 1$	reactants dominate
$\Delta G^0 = 0$	$K = 1$	
$\Delta G^0 = -$ve	$K > 1$	products dominate
$\Delta H = +$ve	$dK/dT = +$ve	more products at higher temperatures
$\Delta H = -$ve	$dK/dT = -$ve	less products at higher temperatures
$\Delta S = +$ve	$dK/dT = +$ve	more products at higher temperatures
$\Delta S = -$ve	$dK/dT = -$ve	less products at higher temperatures
$\Delta V = +$ve	$dK_c/dP = -$ve	less products at higher pressures
$\Delta V = -$ve	$dK_c/dP = +$ve	more products at higher pressures

8.7 The effect of pressure on equilibrium constants

Since K_P can be defined in terms of $\Delta_r G^0$, which in turn can be expressed in terms of the properties of substances in their standard states at one bar pressure, it must itself be independent of pressure. However, this does not mean that the *composition* of the equilibrium mixture will be independent of pressure.

Consider the reaction

$$A_2(g) \rightleftharpoons 2A\,(g)$$

and assume that a fraction α of A_2 (g) is dissociated to produce 2α of A(g), leading to a total number of moles in the mixture of $(1 + \alpha)$. If the total pressure is P, the partial pressure of A_2 is given by

$$\left[\frac{(1-\alpha)}{(1+\alpha)}\right] P \quad \text{and that of A by} \quad \left[\frac{2\alpha}{(1+\alpha)}\right] P.$$

We obtain for the equilibrium constant

$$K_P = \frac{(P_A/\text{bar})^2}{(P_{A_2}/\text{bar})} = \frac{\{[2\alpha/(1+\alpha)][P/\text{bar}]\}^2}{[(1-\alpha)/(1+\alpha)][P/\text{bar}]}$$

$$K_P = \left[\frac{4\alpha^2}{(1-\alpha^2)}\right]\left[\frac{P}{\text{bar}}\right].$$

Since K_P is independent of pressure, the degree of dissociation α must decrease as P increases.

The equilibrium constant of a reaction can be expressed in terms other than pressure. It is frequently expressed in terms of mole fractions, x_i, where $x_i = n_i/N$

and where N is the total number of moles of reactants and product. For the reaction

$$aA + bB \rightleftharpoons lL + mM,$$

we write the equilibrium constant expressed in terms of mole fraction as

$$K_x = \frac{x_L^l x_M^m}{x_A^a x_B^b}.$$

If the reactants and products are perfect gases, then

$$x_i = \frac{P_i}{P} \quad \text{and} \quad K_x \text{ becomes}$$

$$K_x = \frac{(P_L/P)^l (P_M/P)^m}{(P_A/P)^a (P_B/P)^b}$$

$$K_x = \left[\frac{(P_L/\text{bar})^l (P_M/\text{bar})^m}{(P_A/\text{bar})^a (P_B/\text{bar})^b} \right] \left[\frac{P}{\text{bar}} \right]^{(a+b-l-m)}$$

$$K_x = K_P \left(\frac{P}{\text{bar}} \right)^{-\Delta n},$$

where $\Delta n = (l+m-a-b)$, the change in the number of moles of gaseous substances as the reaction proceeds from left to right. Thus, unless $\Delta n = 0$, the mole fractions of the components of the equilibrium mixture will depend on the total pressure, even though K_P does not. If the reaction causes an increase in the number of moles of gas present, an increase in pressure will reduce the mole fraction of products in the final equilibrium mixture. If the reaction causes a decrease in the number of moles of gas present an increase in pressure will increase the mole fraction of products. We can express the dependence of K_x on pressure as

$$\left(\frac{\partial \ln K_x}{\partial P} \right)_T = \left(\frac{\partial \ln K_P}{\partial P} \right)_T - \Delta n \left(\frac{\partial \ln P}{\partial P} \right)_T.$$

Since

$$\left(\frac{\partial \ln P}{\partial P} \right) = \frac{1}{P}$$

and, for perfect gases, $P\Delta_r V = \Delta n RT$ and K_P is independent of pressure, we obtain

$$\left(\frac{\partial \ln K_x}{\partial P} \right)_T = \frac{-\Delta n}{P} = \frac{-\Delta_r V^0}{RT}.$$

This equation applies not only to equilibria involving gases, but also to equilibria in solution and, indeed, to any reaction when the position of equilibrium is expressed in terms of mole fractions rather than partial pressures. In these circumstances, $\Delta_r V^0$ is the volume change accompanying one mole of reaction with all substances in their standard states.

Equilibrium constants can also be expressed in terms of concentrations, $c_i = n_i/V$. In a perfect gas for which $P_i = n_i(RT/V) = c_i RT$ and we obtain

$$K_P = \left(\frac{c_L^l c_M^m}{c_A^a c_B^b}\right)(RT)^{l+m-a-b} - K_c(RT)^{\Delta n} \quad \text{giving}$$

$$K_c = K_P(RT)^{-\Delta n}.$$

Equilibrium constants based on concentrations are most commonly used for reactions taking place in solution. Under these circumstances, the concentrations are expressed in terms of standard states other than that at one bar such as, for example, that of unit molality (1 mol kg^{-1}), or unit molarity (1 mol dm^{-3}).

8.8 Equilibrium calculations using thermodynamic tables

One of the most important aspects of chemical thermodynamics is that it enables us to calculate the behaviour of chemical equilibria from a limited quantity of data that is readily available in tables. We have already used tables to determine the enthalpy changes occurring with the water–gas reaction (Section 2.8)

$$C(s) + H_2O(g) \rightleftharpoons CO(g) + H_2(g).$$

Using the relation

$$\Delta_r H^0 = \Sigma \Delta_f H^0 \text{ (products)} - \Sigma \Delta_f H^0 \text{ (reactants)},$$

we found that the forward reaction is endothermic and that it was necessary to provide heat for the reaction to continue.

Thermodynamic tables can be used more generally to predict the outcome of chemical reactions. The position of equilibrium of any chemical process is determined by the standard Gibbs free energy change, $\Delta_r G^0 = -RT \ln K$ (Section 8.5). The standard free energy change accompanying a reaction at 298 K is given, in terms of the standard free energies of formation of the reactants and products, by

$$\Delta_r G^0 = \Sigma \Delta_f G^0 \text{ (products)} - \Sigma \Delta_f G^0 \text{ (reactants)}.$$

We will take, as an example, the water–gas shift reaction

$$CO(g) + H_2O(g) \rightleftharpoons H_2(g) + CO_2(g).$$

This process reacts the CO produced in the water–gas reaction with steam to produce more hydrogen. It is an important source of hydrogen necessary for the synthesis of ammonia by the Haber process.

From the data given in Appendix 1, for the reactants and products in a gaseous state, we find that, at 298 K

$$\Delta_r G^0 = \Delta_f G^0(CO_2) + \Delta_f G^0(H_2) - \Delta_f G^0(CO) - \Delta_f G^0(H_2O)$$

$$\Delta_r G^0 = -394.36 + 0 + 137.17 + 228.57 = -28.62 \, kJ \, mol^{-1}.$$

Similar calculations using the data from the tables in Appendix 1 give us

$$\Delta_r H^0 = -41.16 \, kJ \, mol^{-1}, \quad \Delta_r S^0 = -42.08 \, J \, K^{-1} \, mol^{-1}$$

and

$$\Delta_r C_P^0 = 3.21 \, J \, K^{-1} \, mol^{-1}.$$

The equilibrium constant for the reaction, K_P, is defined as

$$K_P = \frac{P_{H_2O} P_{CO}}{P_{CO_2} P_{H_2}}$$

and is given at 298 K by

$$\ln K_P = \frac{-\Delta_r G^0}{RT} = \frac{28.62 \times 10^3}{8.314 \times 298} = 11.55$$

$$K_P = 1.04 \times 10^5.$$

The equilibrium position of the reaction is dominated by the products. However, the reaction will not proceed at such a low temperature at a practical rate. We need to consider the characteristics of the reaction at much higher temperatures. The simplest approach that we can make is to assume that $\Delta_r H$ and $\Delta_r S$ are independent of temperature. This is equivalent to assuming that the change in heat capacity accompanying the reaction is zero.

Then, $\Delta_r G^0 = \Delta_r H^0 - T\Delta_r S^0$ and, since $\Delta_r S^0$ is negative as the temperature increases, $\Delta_r G$ will become more positive and the fraction of products in the equilibrium mixture will be reduced. At 1300 K

$$\Delta_r G_T^0 = -41.16 \times 10^3 + 1300 \times 42.08 = 13.54 \, kJ \, mol^{-1}$$

and

$$\ln K_P = \frac{-13.54 \times 10^3}{8.314 \times 1300} = -1.25,$$

giving $K_P = 0.29$. At this temperature, the yield of hydrogen would be very much lower.

We could perform this calculation to obtain K_P from a knowledge of $\Delta_r H^0$ alone using the van't Hoff isochore. (We may drop the subscript P because, as the number of moles of gas on each side of the equation is the same, $K_P = K_c$, see Section 8.7.)

$$\frac{d \ln K}{dT} = \frac{\Delta_r H^0}{RT^2}$$

which, in its integrated form gives (Section 8.6)

$$\ln \left(\frac{K_T}{K_{298}} \right) = \frac{-\Delta_r H^0}{R} \left[\frac{1}{T} - \frac{1}{298} \right]$$

$$\ln \left(\frac{K_{1300}}{K_{298}} \right) = \frac{41.16 \times 10^3}{8.314} \left[\frac{1}{1300} - \frac{1}{298} \right] = -12.81.$$

Thus, $K_{1300} = K_{298} \times 0.27 \times 10^{-5}$ and, since $K_{298} = 1.04 \times 10^5$, again we calculate that $K_{1300} = 0.28$. The experimentally-observed value of K_P at 1300 K is 0.56.

The calculation can be improved if we recognise that $\Delta_r C_P$ is not zero and assume that its value is constant at its value at 298 K which is $3.21 \, \text{J K}^{-1} \, \text{mol}^{-1}$. This will lead to different estimates of both $\Delta_r S$ and $\Delta_r H$ at 1300 K.

$$\Delta_r H_T = \Delta_r H^0 + \Delta_r C_P^0(T - 298) \quad \text{(Section 2.8)}$$

$$\Delta_r H_{1300} = -41.16 + 3.21 \times 1002 \times 10^{-3} = -37.95 \, \text{kJ mol}^{-1}$$

and

$$\Delta_r S_T = \Delta_r S^0 + \Delta_r C_P^0 \ln \left(\frac{T}{298} \right) \quad \text{(Section 7.8)}$$

$$\Delta_r S_{1300} = -42.08 + 3.21 \times \ln \left(\frac{1300}{298} \right) = -37.35 \, \text{J K}^{-1} \, \text{mol}^{-1}.$$

$$\Delta_r G_T = \Delta_r H_T - T\Delta_r S_T$$

$$\Delta_r G_{1300} = -37.95 + 1300 \times 37.35 \times 10^{-3} = 10.61 \, \text{kJ mol}^{-1}.$$

$$\ln K_{1300} = \frac{-\Delta_r G_{1300}}{RT} = \frac{-10.61 \times 10^3}{8.314 \times 1300} = -0.98$$

$$K_{1300} = 0.37$$

Again, a more direct route to carry out essentially the same calculation is available using the integrated form of the van't Hoff isochore.

$$\frac{d \ln (K_T/K_{298})}{dT} = \frac{(\Delta_r H^0 - 298 \times \Delta_r C_P)}{RT^2} + \frac{\Delta_r C_P}{RT}$$

$$\ln (K_T/K_{298}) = \frac{-(\Delta_r H^0 - 298 \times \Delta_r C_P)(1/T - 1/298)}{R} + \frac{\Delta_r C_P \ln (T/298)}{R}$$

$$\ln (K_{1300}/K_{298}) = -12.53$$

$$\frac{K_{1300}}{K_{298}} = 0.36 \times 10^{-5} \quad \text{and} \quad K_{1300} = 0.37$$

The result obtained for the equilibrium constant at 1300 K taking the difference in heat capacity as its value at 298 K is an improvement on the value obtained by assuming that $\Delta_r C_P$ is zero, but still significantly different from the experimental value, 0.56. The results are summarised in Table 8.2.

To improve further our estimate of the equilibrium constant, we must allow for the fact that the difference in the heat capacities of the reactants and products is itself a function of temperature. The data necessary to calculate $\Delta_r C_P$ as a function of temperature are not included in the standard thermodynamic tables we have used so far. However, for common substances in the gaseous state, heat capacities are available, expressed in terms of simple functions of temperature of which the three-term equations

$$C_P = a + bT + cT^{-2} \quad \text{and} \quad \Delta C_P = \Delta a + \Delta bT + \Delta cT^{-2}$$

are commonly employed. Using expressions of this form in conjunction with the integrated form of the van 't Hoff isochore, we can obtain accurate values of

Table 8.2. Summary table: equilibrium constants, K, for the water–gas shift reaction $CO(g) + H_2O(g) \rightleftharpoons H_2(g) + CO_2(g)$.

	Temperature/K		
Method of calculation	298.15	1000	1300
Directly from thermodynamic tables	1.04×10^5		
Van't Hoff isochore, $\Delta_r C_P^0 = 0$		0.86	0.28
Van't Hoff isochore, $\Delta_r C_P^0 = 3.21 \, J \, K^{-1} \, mol^{-1}$		1.1	0.37
Van't Hoff isochore, $\Delta_r C_P^0 = f(T)$		1.44	0.57
From free energy functions		1.40	0.56
From partition functions	0.97×10^5	1.40	0.56
Experimentally observed values		1.41	0.56

equilibrium constants over the temperature range for which the heat capacity data is valid. The results obtained using this approach are much improved and are given in Table 8.2. However, the calculations are laborious and it is much more convenient to use tables of different thermodynamic functions which vary only slowly with temperature. These are called, rather confusingly, free energy functions.

8.9 Equilibrium constants from free energy functions

The property which can be used to calculate equilibrium constants of substances over a wide range of temperature in an efficient manner is the function

$$\frac{(G_T^0 - H_0^0)}{T}.$$

G_T^0 is the Gibbs energy of a substance, under standard conditions, at temperature T, and H_0^0 is the enthalpy of the substance at the absolute zero of temperature. This function, unlike G^0, varies relatively slowly with temperature and can be interpolated even if tabulated at relatively widely spaced temperature intervals. (We should note that H_0^0 is sometimes written E_0^0 since, at the absolute zero of temperature, energy and enthalpy are identical.) For a reaction,

$$\Delta_r G_T^0 = T \Delta_r \left[\frac{(G_T^0 - H_0^0)}{T} \right] + \Delta_r H_0^0.$$

The second term, $\Delta_r H_0^0$, can be identified with the energy or enthalpy change accompanying the reaction at absolute zero and can be identified with the difference in the enthalpies of formation at absolute zero,

$$\Delta_r H_0^0 = [\Delta H_0^0]_{\text{products}} - [\Delta H_0^0]_{\text{reactants}}.$$

The values of ΔH_0^0 for substances are included in tables of free energy functions to enable $\Delta_r H_0^0$ for a reaction to be calculated. At 1300 K,

$$\Delta_r G_T^0 = T \Delta_r \left[\frac{(G^0 - H_0^0)}{T} \right] + \Delta_r H_0^0$$

$$\Delta_r G_{1300}^0 = 1300 \times 35.90 \times 10^{-3} + (-40.42) = 6.25 \text{ kJ mol}^{-1}$$

Table 8.3. Calculation of equilibrium constants for the water–gas shift reaction from free energy functions.

		$-[\Delta_r(G_T^0 - H_0^0)/T]$ $JK^{-1}\,mol^{-1}$		$-\Delta_r H_0^0/$ $kJ\,mol^{-1}$
T/K	298.16	1000	1300	
CO_2	182.26	226.40	238.15	393.17
H_2	102.17	136.98	144.72	0.000
CO	168.41	204.05	212.54	113.81
H_2O	155.52	196.69	206.23	238.94
$[\Delta_r(G_T^0 - H_0^0)/T]/J\,mol^{-1}$	39.50	37.36	35.90	—
$\Delta_r H_0^0$	—	—	—	40.42
$-\Delta_r G_T^0$	28.64	3.06	-6.25	—
K_{calc}	1.04×10^5	1.40	0.56	
K_{obs}	—	1.41	0.56	

$$K = \exp\left(\frac{-\Delta_r G_{1300}^0}{RT}\right) = \exp\left(\frac{-6.25 \times 10^3}{8.314 \times 1300}\right)$$

$$K = 0.56.$$

This value is identical with the experimentally-observed value and illustrates the superiority of this method of calculation. The results are summarised in Table 8.3.

8.10 Equilibrium constants and partition functions

An alternative to the thermodynamic routes for the calculation of equilibrium constants of chemical reactions is via partition functions (introduced in Section 7.9). This requires knowledge of the specific properties of the molecules involved. We have established that the partition functions are measures of the effective number of energy levels available to the system at a given temperature. They can be calculated for each molecule from the mass, dimensions and vibrational frequencies. The thermodynamic properties of the molecules can then be calculated from knowledge of the molecular partition function, z, and the temperature and volume derivatives (Table 7.1).

$$G^0 = -RT \ln\left(\frac{z^0}{N_A}\right),$$

where N_A is the Avogadro constant and $\Delta_r G^0 = -RT \ln K_P$, where z^0 is the molecular partition function evaluated at the standard state of one bar pressure.

We obtain for the general reaction

$$aA + bB \rightleftharpoons lL + mM,$$

$$K_P = \left[\frac{\left(z_L^0/N_A \right)^l \left(z_M^0/N_A \right)^m}{\left(z_A^0/N_A \right)^a \left(z_B^0/N_A \right)^b} \right] \exp \left(\frac{-\Delta_r E_0^0}{RT} \right).$$

$\Delta_r E_0^0$ is the difference between the energies of the products and reactants in the lowest vibrational-rotational levels. It is necessary to include this term so that all the energy levels are referred to the same zero of energy.

If we apply this to the water–gas shift reaction

$$CO + H_2O \rightleftharpoons CO_2 + H_2$$

and, noting again that, as the number of moles of gases is unchanged, $K_P = K_c$, we can write

$$K = \frac{z^0(CO_2)z^0(H_2)}{z^0(H_2O)z^0(CO)} \exp \left(\frac{-\Delta_r E_0^0}{RT} \right),$$

where

$$\Delta_r E_0^0 = -E_0^0(CO_2) - E_0^0(H_2) + E_0^0(CO) + E_0^0(H_2O).$$

We make the assumption that the total molecular energy can be expressed in terms of independent contributions from the electronic, translational, rotational and vibrational modes. For each molecule,

$$E = E_{trans} + E_{vib} + E_{rot} + E_{elec}$$

and

$$z = z_{trans} \, z_{vib} \, z_{rot} \, z_{elec}.$$

We also assume that the rotation of molecules may be represented by rigid rotors and the vibrational modes by harmonic oscillators. (See Section 7.10 for the relevant background to the equations employed in the calculations below.) $z_{elec} = g_E$, the degeneracy of the ground electronic state. For all substances involved in the water–gas shift reaction, $z_{elec} = 1$.

The expression for the translational partition function is

$$z_{trans} = \left(\frac{2\pi mkT}{h^2} \right)^{\frac{3}{2}} V.$$

For the linear molecules involved in the reaction,

$$z_{rot} = \left(\frac{8\pi^2 IkT}{h^2 \sigma} \right),$$

where I is the moment of inertia and σ the symmetry factor. For a non-linear molecule, in our current example H_2O, the rotational partition function is given by

$$z_{\text{rot}} = \frac{\pi^{\frac{1}{2}}(8\pi^2 kT)^{\frac{3}{2}}(I_x I_y I_z)^{\frac{1}{2}}}{h^3\sigma}.$$

The partition function of each vibrational mode is

$$z_{\text{vib}} = \left[1 - \exp\left(\frac{-h\upsilon}{kT}\right)\right]^{-1}.$$

The molecular parameters for the molecules involved in the water–gas shift reaction are given in Table 8.4. The contribution from the translational partition functions of the molecules, which we will write as Z_{trans}, can be expressed

$$Z_{\text{trans}} = \left(\frac{m_{CO_2}m_{H_2}}{m_{H_2O}m_{CO}}\right)^{\frac{3}{2}}$$

and depends only on the molecular masses, since all the other factors in the equation for the translation partition function cancel out, giving

$$Z_{\text{trans}} = \left(\frac{44.01 \times 2.016}{18.02 \times 28.01}\right)^{\frac{3}{2}} = 0.0737.$$

We would have similar simplicity for the contribution from the rotational partition functions (which would depend only on the ratio of the moments of inertia) were it not for the complications introduced by the fact that water is a non-linear molecule.

Table 8.4. Molecular properties of the calculation of the equilibrium constant for the water–gas shift reaction.

	CO_2	H_2	H_2O	CO	Δ
E_0^0 /kJ mol^{-1}	1596	432.1	917.6	1070	−40.50
Molar mass	46.00	2.016	18.02	28.01	—
Bond length/Å	1.160	0.741	0.957	1.128	
Bond angle/degrees	—	—	104.47	—	—
Vibration frequencies/cm^{-1}	2335	4319	3725	2156	—
	663		3586		
	1313		1591		
Moment of inertia/amu Å^2	43.0	0.277	1.13*	8.73	—
Symmetry number	2	2	2	1	

$*(I_x I_y I_z)/\text{amu}^{3/2}\text{Å}^3$, $\qquad\qquad\qquad\qquad\qquad\qquad 1\,\text{Å} = 10^{-10}\text{m}$

Substituting in the expressions for the rotational partition functions, we obtain

$$Z_{rot} = \left[\frac{I_{CO_2} I_{H_2}}{(I_x I_y I_z)_{H_2O}^{\frac{1}{2}} I_{CO}} \right] \left[\frac{\sigma_{CO}\sigma_{H_2O}}{\sigma_{CO_2}\sigma_{H_2}} \right] \left[\frac{8\pi^3 kT}{h^2} \right]^{-1/2}.$$

If we express the moments of inertia in amu and nm, this reduces to

$$Z_{rot} = \left[\frac{I_{CO_2} I_{H_2}}{(I_x I_y I_z)_{H_2O}^{\frac{1}{2}} \cdot I_{CO}} \right] \left[\frac{\sigma_{CO}\sigma_{H_2O}}{\sigma_{CO_2}\sigma_{H_2}} \right] \left[\frac{0.2774}{T^{\frac{1}{2}}} \right],$$

and, substituting for the values of the moments of inertia and symmetry numbers, we obtain

$$Z_{rot} = \left[\frac{0.43 \times 0.00277}{1.13 \times 10^{-3} \times 0.0873} \right] \left[\frac{1}{2} \right] \left[\frac{0.2774}{T^{\frac{1}{2}}} \right] = \frac{1.679}{T^{\frac{1}{2}}}.$$

The contributions from the vibrational modes of the molecules

$$Z_{vib} = \Pi \left[1 - \exp\left(\frac{-h\nu}{kT} \right) \right]^{-1}$$

are evaluated by a direct calculation for each vibrational mode or, more conveniently, by reference to Table 7.5. $\Delta_r E_0^0$ for this reaction has the value $-40.50\,kJ\,mol^{-1}$ (Table 8.4). The contributions from the various modes are set out in Table 8.5 and the results, together with the results of the other calculations carried out previously, are summarised in Table 8.2. At 1300 K, the equilibrium constant is calculated to be 0.56, which is in excellent agreement with the experimental value. Inaccuracies in these calculations arise from the assumption of rigid rotor and harmonic oscillator models and can be corrected for if detailed spectroscopic information is available.

Table 8.5. Contribution to the partition functions from different modes for the equilibrium constant of the water–gas shift reaction.

Contribution to z	298.15 K	1000 K	1300 K
Translation	0.0737	0.0737	0.0737
Rotation	0.0973	0.0531	0.0466
Vibration	1.088	2.747	3.867
$\exp(-\Delta E_0^0/RT)$	1.25×10^7	130.5	42.4
K_{calc}	0.97×10^5	1.40	0.56
K_{obs}	—	1.41	0.56

8.11 Summary of the basic equations of chemical thermodynamics

For a system at constant temperature and pressure which does only PV work, the position of equilibrium can be defined in terms of the minimum Gibbs free energy G, where $dG = 0$. The free energy is defined as $G = H - TS$. At low temperatures, the tendency is to minimise the energy and the most stable configurations are characterised by large negative values of the enthalpy, H. At high temperatures, the tendency to maximum entropy, S becomes most important. The equation

$$dG = VdP - SdT + \Sigma\mu_i dn_i,$$

where μ_i is the chemical potential of substance i, is often called the *fundamental equation* of chemical thermodynamics (Sections 8.3 and 8.4). It describes how the Gibbs free energy varies with temperature, pressure and the chemical composition of the system.

The important equations that occur in chemical thermodynamics are relatively few and are of extremely wide applicability. Those we have obtained for equilibrium in chemical reactions are quite general and apply to all equilibria.

In summary:

(i) We may express the *chemical potential* of any substance i by

$$\mu_i = \mu_i^0 + RT \ln \left(\frac{c_i}{c^0}\right),$$

where c_i is a measure of the quantity of component i relative to that in a standard state c^0. c could, for example, represent the partial pressure, concentration or mole fraction of i (Section 8.4).

(ii) For a reaction, under conditions of constant temperature and pressure, we may write (Section 8.5)

$$\Delta G' = \left(\frac{\partial G}{\partial \xi}\right)_{T,P} = \Sigma v_i \mu_i,$$

where v_i are the stoichiometric coefficients,
and, since

$$\mu_i = \mu_i^0 + RT \ln \left(\frac{c_i}{c^0}\right),$$

we obtain

$$\Delta G' = \left(\frac{\partial G}{\partial \xi}\right)_{T,P} = \Delta G^0 + RT \ln \prod \left(\frac{c_i}{c^0}\right)^{v_i}.$$

ΔG^0 is the standard Gibbs free energy change for one mole of reaction for the equation which describes the equilibrium (that is, the reaction with all substances in their standard states).

(iii) For any system at equilibrium, since

$$\left(\frac{\partial G}{\partial \xi} \right)_{T,P} = 0,$$

we may write

$$\Delta G^0 = -RT \ln K,$$

where K characterises the equilibrium position in terms of the amount of material in the equilibrium mixture. Thus, K could be the equilibrium constant of the chemical reaction, a solubility or a vapour pressure.

(iv) The temperature variation of any equilibrium is given by equations of the form

$$\left(\frac{\partial \ln K}{\partial T} \right)_P = \frac{\Delta H^0}{RT^2},$$

where ΔH^0 is the standard enthalpy change for the reaction (Section 8.6).

(v) The effect of pressure on the position of equilibrium is given by equations of the type

$$\left(\frac{\partial \ln K_x}{\partial P} \right)_T = \frac{-\Delta V^0}{RT},$$

where ΔV^0 is the volume change accompanying a mole of reaction under standard conditions. (Remember that K_P is independent of pressure; Section 8.7.)

We will meet equations of this general form in a number of different circumstances in later chapters. The changes in equilibrium constants in response to changes of thermodynamic functions are summarised in Table 8.1.

Problems

(1) Calculate the standard changes in free energy and enthalpy and evaluate the equilibrium constant for the reaction

$$2SO_2 + O_2 \rightleftharpoons 2SO_3 \text{ at } 298 \text{ K}$$

using the data from Appendix 1.

(2) For the equilibrium $n\text{-}C_4H_{10} \rightleftharpoons i\text{-}C_4H_{10}$, use the data for the Gibbs free energy changes of formation of the two isomers given in Appendix 1 to calculate the composition of the equilibrium mixture at 298 K.

(3) For the equilibrium $H_2 + I_2 \rightleftharpoons 2HI$, at 755 K, the equilibrium partial pressures for hydrogen and iodine are 0.042 atm and for hydrogen iodide 0.29 atm. At 667 K, the equilibrium constant, $K = 61$. Calculate the equilibrium constant at 755 K and estimate the standard enthalpy change accompanying the reaction.

(4) For the equilibrium $N_2O_4 \rightleftharpoons 2NO_2$, the equilibrium constant at 298 K is 0.14 and, at 318 K, 0.67. Calculate the standard enthalpy change and entropy change for the reaction.

(5) At 10 bar pressure and temperature 750 K, the partial pressures of NH_3, N_2 and H_2 are 0.20, 2.4 and 7.4 bar, respectively. Calculate the equilibrium constant and the standard free energy of the reaction

$$N_2 + 3H_2 \rightleftharpoons 2NH_3$$

at this temperature.

9

The States of Matter

9.1 Gases, liquids and solids

Matter can exist in many forms but, most commonly, we identify three distinct states: gas, liquid and solid. The state with the lowest Gibbs free energy is the stable form of matter at any particular temperature. At low temperatures, the solid with the most negative energy is the most stable form. At high temperatures, the gaseous state with the maximum randomness prevails. At intermediate temperatures, the liquid state has the lowest free energy. If the Gibbs free energy is plotted against the temperature at constant pressure, since $dG = VdP - SdT$, the slopes of the lines are $(\partial G/\partial T)_P = -S$. The gaseous phase with the highest entropy has the largest negative slope and will have the lowest free energy at high temperatures (Fig. 9.1). The solid phase with the lowest slope has the lowest free energy at low temperatures. The transition from one phase to another occurs where the lines intersect and where the free energies are equal.

Then, $\Delta G = 0$ and $\Delta H = T\Delta S$, giving, at the melting point of the solid,

$$\Delta_{fus} S = \frac{\Delta_{fus} H}{T_{fus}}$$

and, at the boiling point of the liquid,

$$\Delta_{vap} S = \frac{\Delta_{vap} H}{T_{vap}}.$$

We can represent the equilibrium of the phases as a function of temperature and pressure as shown in Fig. 9.2 for the phase equilibria in water. Such diagrams are called *phase diagrams*. The lines represent the pressures and temperatures at which two phases are in equilibrium. The line AB represents the vapour pressure of liquid water and AC gives the vapour pressure of ice. AD is the melting curve on which the solid and liquid are at equilibrium. The point where all three curves meet, A, is called the *triple point*, which, for water, is at 273.16 K and 0.61 kPa (4.58 mm Hg) pressure, the only pressure and temperature at which the three phases can co-exist in equilibrium.

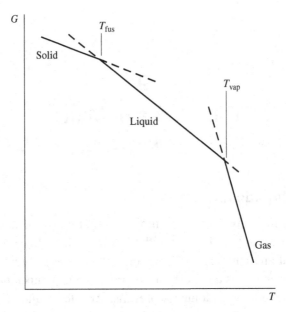

Fig. 9.1. Gibbs free energy as a function of temperature for a pure substance at constant pressure. The slopes of the lines are determined by the entropy of the phases as $(\partial G/\partial T)_P = -S$.

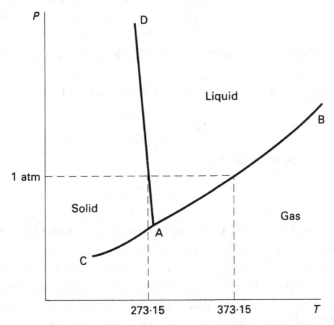

Fig. 9.2. Phase diagram for water (schematic: not to scale). The line AD represents the equilibrium between solid and liquid and is comparatively insensitive to changes in pressure. The line AB represents the equilibrium between liquid and vapour and is considerably more sensitive to pressure because of the large difference in the molar volumes of the two phases. The triple point, A, is fixed and independent of pressure and temperature.

The free energy of each of the phases as a function of pressure is determined by $(\partial G/\partial P)_T = V$ and the equilibrium between the phases as a function of pressure depends on their difference in molar volume ΔV. The volume change from liquid to vapour is large and the boundary is very sensitive to pressure. By contrast, the volume difference between liquid water and ice is relatively small and the melting curve is much less sensitive to changes in pressure.

9.2 The thermodynamics of phase changes

Consider two phases, for example a solid and liquid, in equilibrium at a temperature T and pressure P, when (see Section 8.4)

$$G(l) = G(s).$$

If we now alter the conditions to $T + dT$ and $P + dP$, then

$$dG(l) = V(l)dP - S(l)dT$$

and

$$dG(s) = V(s)dP - S(s)dT.$$

If the two phases remain in equilibrium, then $dG(l) = dG(s)$ and

$$V(l)dP + S(l)dT = V(s)dP + S(s)dT.$$

Thus,

$$\frac{dP}{dT} = \frac{[S(l) - S(s)]}{[V(l) - V(s)]} = \frac{\Delta_{fus}S}{\Delta_{fus}V}.$$

Since the two phases are in equilibrium, $\Delta_{fus}G = 0$ and $\Delta_{fus}S = \Delta_{fus}H/T$. Here, T is the melting point of the liquid at the pressure under consideration, giving

$$\left(\frac{dP}{dT}\right)_{equil} = \frac{\Delta_{fus}H}{T\Delta_{fus}V}.$$

In general, for the equilibrium between any two phases, we have

$$\left(\frac{dP}{dT}\right)_{equil} = \frac{\Delta H}{T\Delta V}.$$

This equation, which is exact, is known as the *Clapeyron* equation.

When applied to the process of vaporization, we can write $\Delta_{vap}V = V(g) - V(l)$. At room temperature and one bar pressure, we have, for one mole of substance,

$V(g) \approx 24000 \, \text{cm}^3$ and $V(l) \approx 100 \, \text{cm}^3$. If we ignore $V(l)$ and replace $V(g)$ by the value for one mole of a perfect gas, $V(g) = RT/P$, we obtain

$$\left(\frac{dP}{dT}\right) = \frac{\Delta_{\text{vap}} H P}{RT^2}$$

and

$$\frac{d \ln P}{dT} = \frac{\Delta_{\text{vap}} H}{RT^2}.$$

This equation is called the *Clausius–Clapeyron* equation and, unlike the Clapeyron equation, is not exact. It is, however, extremely useful in providing a method of estimating $\Delta_{\text{vap}} H$. If $\Delta_{\text{vap}} H$ is assumed to be independent of temperature, we can integrate the equation to obtain

$$\ln P = \frac{-\Delta_{\text{vap}} H}{RT} + \text{const.}$$

If we plot the logarithm of the vapour pressure of the liquid, $\ln P$, against reciprocal temperature, we obtain a linear plot with a slope of $-\Delta_{\text{vap}} H/R$ (or $\Delta_{\text{vap}} H/2.303R$ if $\log P$ is plotted); see Fig. 9.3.

If we assume that $G(l)$ is independent of pressure and $G(l) = G^0(l)$, the value at 1 bar, when one mole of liquid vaporizes to the perfect gas state with vapour

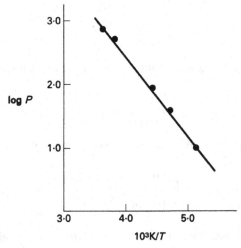

Fig. 9.3. The logarithm of the vapour pressure of liquid n-butane as a function of reciprocal temperature.

pressure P, the Gibbs free energy change is given by

$$\Delta_{\text{vap}} G = G(\text{g}) - G(\text{l}) = G^0(\text{g}) + RT \ln \left(\frac{P}{\text{bar}} \right) - G^0(\text{l}).$$

If the gas and vapour are in equilibrium, $\Delta_{\text{vap}} G = 0$ and

$$\Delta_{\text{vap}} G^0 = -RT \ln \left(\frac{P}{\text{bar}} \right).$$

Thus, the vapour pressure of a liquid is determined by the free energy change when one mole of liquid is vaporized to produce one mole of gas *at one bar pressure*, the standard state.

9.3 Intermolecular energy

The phase in which a substance exists under any given conditions depends on how its molecules interact with one another. Gases, under normal conditions, often follow, to a close approximation, the perfect gas equation, $PV = nRT$. As we saw in Section 2.2, this behaviour arises from the fact that the molecules exert no forces on one another and can be regarded as point masses with zero volume. However, if the molecules do interact, deviations from the perfect gas equation occur and, at lower temperatures, the liquid and solid states become the stable forms of matter. The simplest model of intermolecular interaction is that where the molecules can be regarded as hard spheres which do not attract each other, but do occupy space. Using computer simulations, it was found, surprisingly, that a substance composed of such molecules, in which attractive forces can play no role, can show two distinct phases, solid and fluid. A transition can occur which is very similar to the melting process in real substances (Fig. 9.4), emphasising the importance of the increased disorder of the fluid in driving the melting process.

For real substances, the molecules possess both size, which is determined by repulsive forces, and attractive forces as illustrated in Fig. 9.5, which shows how the intermolecular energy of two spherically symmetric, but more realistic, molecules varies as a function of the separation. The force between the molecules is the slope of the energy versus separation curve, $F(r) = -(\text{d}u/\text{d}r)$ and $u(r) = \int_r^\infty F(r)\text{d}r$. (We identify the intermolecular potential energy by lower case u in order to avoid confusion with the internal energy or volume when, as is often the case, it is represented by U or V.) At short range, steep repulsive forces arise from the overlap of the electronic distributions in the molecules. At long range, the forces are attractive in nature. The forces are balanced where $(\text{d}u/\text{d}r) = 0$ at the minimum of the energy curve. At this point, at a separation of r_m, the energy of attraction has its maximum value at $-\varepsilon$.

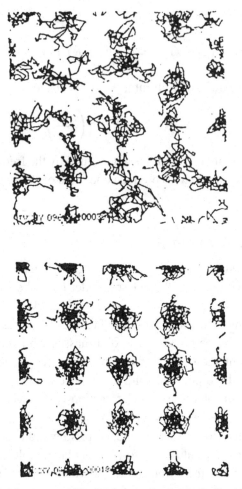

Fig. 9.4. Traces made by the centres of hard sphere molecules in the liquid phase (upper diagram) and the solid phase (lower diagram). The transition from liquid to solid takes place suddenly, even though the system remains under unchanged conditions.

We may estimate the order of magnitude of these forces by noting that the entropy of vaporization at the boiling point for typical nonpolar liquids is given by Trouton's rule, $\Delta_{\text{vap}}S = \Delta_{\text{vap}}H/T_{\text{vap}} \approx 85\,\text{J K}^{-1}\,\text{mol}^{-1} \approx 10\,\text{R}$. If we assume that each molecule in the liquid is surrounded by ten nearest neighbours, each approximately at the distance of the minimum energy $-\varepsilon$, then, for one mole, $\Delta_{\text{vap}}H \approx 10N_A\varepsilon/2$ (the factor of two is to prevent us counting the interaction of each pair of molecules twice over). Thus, equating the two expressions for $\Delta_{\text{vap}}H$, we obtain $N_A\varepsilon \approx RT_{\text{vap}}$ and, since $k = R/N_A$, $\varepsilon \approx kT_{\text{vap}}$. For simple molecules, such as argon or nitrogen, which boil at temperatures well below room temperature, the intermolecular energy for a mole of substance is $N_A\varepsilon \approx 1\,\text{kJ mol}^{-1}$, whereas for substances that boil just

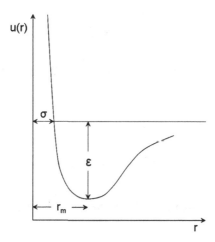

Fig. 9.5. Intermolecular potential energy as a function of separation.

above room temperature, such as benzene, the intermolecular energy is of the order of $3\,\text{kJ}\,\text{mol}^{-1}$.

9.4 The origins of intermolecular energy

When atoms and molecules approach within short distances of each other, unless a chemical bond is formed, repulsion occurs. The *repulsive energy* arises partly from the interactions of the incompletely screened nuclei and partly from the direct repulsion of the electrons. We have seen in Chapter 5 that the valence bond and molecular orbital theories give an account of the repulsion between atoms when no bonding occurs, but neither offers a route to the quantitative evaluation of the forces for systems containing more than a few electrons. To a large extent, our knowledge of the repulsive forces between molecules is based on experimental studies. It has been found that the forces are best represented by a series of exponential terms.

The sources of *intermolecular attraction* between non-ionic species, van der Waals forces, are threefold.

(i) *Electrostatic energy*

The intermolecular potential energy, identified as electrostatic energy, arises from the fact that molecules with distributed electrical charges will give rise to an electrostatic potential and associated electrical field. Consider a molecule that is represented by a linear charge distribution consisting of two charges q_1 and q_2 located on the z axis at coordinates $-z_1$ and z_2 (Fig. 9.6). At point P, a distance r from the origin and at

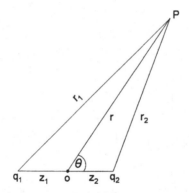

Fig. 9.6. A linear charge distribution.

angle θ to the z axis, the electrostatic potential is given by

$$\Phi(4\pi\varepsilon_0) = \left(\frac{q_1}{r_1} + \frac{q_2}{r_2}\right).$$

r_1 and r_2 can be expressed in terms of r, z_1, z_2 and θ, giving

$$\Phi(4\pi\varepsilon_0) = \frac{q_1}{(r^2 + z_1^2 + 2z_1 r \cos\theta)^{1/2}} + \frac{q_2}{(r^2 + z_2^2 - 2z_2 r \cos\theta)^{1/2}}.$$

If r is significantly greater than z_1 and z_2, we can expand the denominators of the two terms in powers of z_1/r and z_2/r to obtain

$$\Phi(4\pi\varepsilon_0) = \frac{(q_1 + q_2)}{r} + \frac{(q_2 z_2 - q_1 z_1)\cos\theta}{r^2} + \frac{(q_1 z_1^2 + q_2 z_2^2)(3\cos^2\theta - 1)}{2r^3} + \cdots$$

We can write this expression in the form

$$\Phi(4\pi\varepsilon_0) = \frac{Q}{r} + \frac{\mu\cos\theta}{r^2} + \frac{\Theta(3\cos^2\theta - 1)}{2r^3} + \cdots,$$

where Q is the total charge, $\mu = (q_2 z_2 - q_1 z_1)$ is the dipole moment and $\Theta = (q_1 z_1^2 + q_2 z_2^2)$ is the quadrupole moment of the charge distribution (Section 2.10). (We have restricted a consideration to a linear charge array since to consider a three-dimensional case would require us to recognise that the dipole moment is a vector and the quadrupole moment a tensor, and lead us into a much more difficult analysis.)

If we have two linear molecules labelled A and B with no overall electrical charge oriented as in Fig. 9.7, the energy of interaction of the two molecules can be

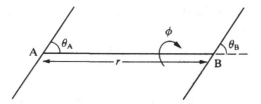

Fig. 9.7. The angles defining the relative orientations of two linear charge distributions.

written in terms of dipole, quadrupole and higher moments

$$u_{AB}(4\pi\varepsilon_0) = \frac{\mu_A\mu_B f_1(\theta_A, \theta_B, \varphi)}{r^3} + \frac{(\mu_A\Theta_B + \mu_B\Theta_A)f_2(\theta_A, \theta_B, \varphi)}{r^4}$$

$$+ \frac{\Theta_A\Theta_B f_3(\theta_A, \theta_B, \varphi)}{r^5} + \cdots .$$

We see that the energy falls off more rapidly for the contributions from higher multiple moments. The first term representing the interaction of dipolar molecules is

$$u_{AB}(4\pi\varepsilon_0) = \frac{-\mu_A\mu_B(2\cos\theta_A\cos\theta_B - \sin\theta_A\sin\theta_B\cos\varphi)}{r^3}.$$

When the dipoles are oriented in the same direction in both molecules, this becomes $u_{AB} = -2\mu_A\mu_B/(4\pi\varepsilon_0)r^3$. When the dipoles are parallel to each other but point in opposite directions, the energy has the same magnitude but is of the opposite sign. Thus, the energy depends strongly on orientation and we can show, by considering the symmetry, that if the energy is simply averaged over all orientations, weighting them equally, the net energy is zero. This might suggest that dipolar molecules would not attract each other. However, a net attraction between two dipolar molecules arises because the configurations with negative energy are weighted more heavily by their Boltzmann factors. The electrostatic energy of interaction averaged over all orientations weighted by the Boltzmann factors becomes

$$\langle u_{\text{elec}} \rangle = \frac{\int u(\omega)\exp[-u(\omega)/kT]d\omega}{\int \exp[-u(\omega)/kT]d\omega},$$

where ω represents the mutual orientations of the molecules. If $u(\omega) \ll kT$, we can expand the exponential terms, since $\exp(-x) = 1 - x + x^2 \cdots$, to obtain

$$\langle u_{\text{elec}} \rangle = \int u(\omega)d\omega - \int [u(\omega)]^2/kTd\omega \ldots.$$

Substituting the energy of interaction of two dipolar molecules for $u(\omega)$ and integrating over all orientations, we find that the first term averages to zero and the leading contribution to the overall energy of interaction becomes

$$\langle u_{\text{elec}} \rangle = \frac{-2\mu_A^2 \mu_B^2}{3r^6 kT(4\pi\varepsilon_0)^2}.$$

Thus, the dipole contribution to the intermolecular energy depends on the product of the squares of the dipole moments of molecules and is inversely proportional to temperature and to the separation, r, to the sixth power.

If the molecules have quadrupole moments, then a similar averaging process leads to the result

$$\langle u_{\text{elec}} \rangle = \frac{-14\Theta_A^2 \Theta_B^2}{5(4\pi\varepsilon_0)^2 kTr^{10}}.$$

(ii) *Induction energy*

When a molecule is placed in an electric field, F, the positive and negative charges within the molecule redistribute. This *polarisation* produces an induced dipole moment that is proportional to the field: $\mu_{\text{ind}} = \alpha F$, where α is the polarisability of the molecule. This induced dipole can then interact with the dipolar molecule that gave rise to the field, to produce a net attraction (after appropriate averaging over all angles) of

$$\langle u_{\text{ind}} \rangle = \frac{-\mu^2 \alpha}{(4\pi\varepsilon_0)^2 r^6}.$$

If the interacting molecules are both dipolar and polarisable, then we have, for the total energy of interaction due to induction,

$$\langle u_{\text{ind}} \rangle = \frac{-2\mu^2 \alpha}{(4\pi\varepsilon_0)^2 r^6}.$$

In this case, the leading term does not average to zero, as is the case for dipolar interactions, and, consequently, the energy of interaction does not depend on temperature.

(iii) *Dispersion energy*

The two contributions to intermolecular energy, the forces considered above, are easily understood in terms of classical physics. However, even molecules that have no permanent dipole (or higher multipole moments) and where induction effects are absent still show long-range intermolecular attraction. This most important source of intermolecular energy was first evaluated by Fritz London in 1930 using quantum mechanical arguments.

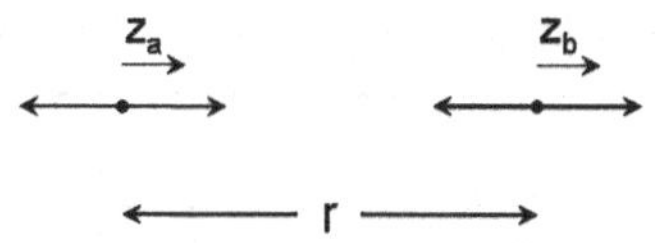

Fig. 9.8. The interaction of two sets of oscillating electric charges: the model for calculating the magnitude of dispersion forces.

We consider two sets of oscillating charges as illustrated in Fig. 9.8. The oscillators each have a force constant k and, when separate, each will have the zero-point energy $\frac{1}{2}h\upsilon$. If we bring the two oscillators together, they will interact and an additional electrostatic contribution to the overall energy will be present, given by

$$\frac{-z_a z_b Q^2}{(4\pi\varepsilon_0)r^3},$$

where z_a and z_b are the amplitudes of the two oscillators and Q the charges. When we solve the Schrödinger equation with this extra term, we find that the additional energy is given by

$$u = \frac{-Q^4 h\upsilon}{2(4\pi\varepsilon_0)^2 r^6 k^2}$$

and the zero-point energy of the system will differ from the zero-point energy of the separated oscillators.

If we apply this model to molecules in which the electrons can be regarded as 'oscillating' about the nuclei, we can obtain an expression for the dispersion energy. The force constant k can be related to the polarisability of the molecules, α, by $\alpha = Q^2/k$. London assumed that $h\upsilon$ could be identified with the ionization energy of the molecule, I, and obtained the equation

$$\langle u_{dis} \rangle = \frac{-3\alpha^2 I}{4(4\pi\varepsilon_0)^2 r^6} = \frac{C_6}{r^6}.$$

This leading term arises from the instantaneous dipole-induced dipole interactions. In fact, higher terms in r^{-8} and r^{-10} etc. also contribute to the dispersion energy, but the r^{-6} term dominates.

The *total attractive energy* between molecules is dominated by the sum of each of these contributions,

$$u = u(\text{electrostatic}) + u(\text{induction}) + u(\text{dispersion})$$

and, for the interactions of two identical molecules, we obtain

$$u = -\left[\frac{2\mu^4}{3kT(4\pi\varepsilon_0)^2} + \frac{2\mu^2\alpha}{(4\pi\varepsilon_0)^2} + \frac{3\alpha^2 I}{4(4\pi\varepsilon_0)^2} \right] r^{-6}.$$

Table 9.1. Relative contributions to intermolecular energy (arbitrary units).

	μ/D	U_{elec}	U_{ind}	U_{disp}
Coefficients of $r^{-6} \times (4\pi\varepsilon_0)^2$		$2\mu^4/3kT$	$2\mu^2\alpha$	$3\alpha^2 I/4$
Ar	0	0	0	50
Xe	0	0	0	209
HCl	1.03	17	6	150
H_2O	1.84	180	10	61

The relative importance of these contributions for a number of simple molecules is given in Table 9.1. For all but small, strongly dipolar molecules, the dispersion forces tend to dominate.

For simple nonpolar molecules, in particular monatomic species, the behaviour of the intermolecular energy, u, as a function of separation, r, is often represented by the Lennard–Jones function

$$u(r) = 4\varepsilon \left[\left(\frac{\sigma}{r}\right)^{12} - \left(\frac{\sigma}{r}\right)^{6} \right],$$

where σ is the collision diameter at $u(r) = 0$ and ε is the maximum attractive energy (well depth); Fig. 9.5. This potential energy function gives the qualitative features required but provides a poor quantitative representation of the intermolecular forces of even the simplest molecules. One source of the failure is that the repulsive forces are not well represented by a single power. To reproduce accurately the behaviour of the forces would require a number of exponential terms of the form $u = \exp(-ar/r_m)$. For example, in argon, one of the simplest and most studied substances, the collision diameter $\sigma = 335$ pm and the maximum (negative) energy $\varepsilon/k = 142$ K, whereas the traditionally employed Lennard–Jones potential gives values of $\sigma = 340$ pm and $\varepsilon/k = 120$ K.

When calculating the properties of condensed phases, it is usual to assume that the total energy is made up of the sum of all the energy arising from the interactions of pairs of molecules,

$$u_N = \frac{1}{2} \sum \sum u_{ij},$$

the factor of two ensuring that the interactions are not counted twice. This assumption of *pairwise additivity* is not accurate when the presence of a third molecule can perturb the interaction of two other molecules. This occurs both with induction forces and dispersion forces. The leading correction term for dispersion forces when three molecules interact, the triple dipole contribution, was first formulated by Axilrod and

Teller. When the three molecules form a linear array, dispersion forces are enhanced, and when they form a triangle, forces are weakened. The overall effect in a solid or liquid is to reduce the intermolecular energy below that calculated on the assumption of pairwise additivity.

9.5 Gas imperfection

One clear manifestation of intermolecular forces is the manner in which gases deviate from the perfect gas equation of state. These deviations may be investigated by studying the behaviour of the *compression factor*, $Z = PV/nRT$, as a function of pressure and temperature (Fig. 9.9). For perfect gases, Z is unity at all temperatures and pressures. For real gases, at low temperatures, Z typically falls below unity as the pressure is raised but, as the pressure is increased further, it rises to values greater than unity. At high temperatures, the initial fall does not occur and Z increases monotonically as the pressure is raised. At a particular temperature, the Boyle temperature, T_B, the initial slope is zero. The values of $Z < 1$ are caused by the attractive forces dominating, pulling the molecules together, making the volume occupied by a fixed quantity of gas less than would be occupied by a perfect gas. Z becomes greater than unity when the finite size of the molecules (repulsive forces) dominates. Under these circumstances, the molecules take up room and reduce the

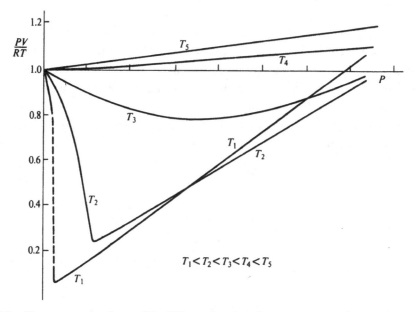

Fig. 9.9. The compression factor, PV_m/RT as a function of pressure at several temperatures. T_2 is the critical temperature and T_4 is the Boyle temperature.

effective volume in which the molecules can move, leading to an increase in the pressure.

This behaviour was first explained by van der Waals in 1873. He represented gas molecules as hard spheres of finite size which could attract one another. The expression he obtained for the equation of state of a real gas, the *van der Waals equation of state*, is

$$\left(P + \frac{a}{V_m^2}\right)(V_m - b) = RT.$$

$V_m = V/n$ is the molar volume of the gas and b is defined as the volume unavailable to the centre of the molecules due to the nonzero size of the molecules. This is four times the actual volume of the atoms or molecules (Fig. 9.10),

$$b = \frac{2\pi N_A d^3}{3},$$

where d is the diameter of the molecules. The parameter a reflects the effect of attraction between the molecules.

We can write the equation in the form

$$\frac{PV_m}{RT} = 1 + \frac{(b - a/RT)}{V_m} + \frac{b^2}{V_m^2} + \frac{b^3}{V_m^3}\cdots.$$

In this form, we can see that, at low temperatures, the term a/RT will dominate and lead to negative deviations from perfection. At high temperatures, the term in b dominates and leads to positive deviations. At the Boyle temperature, the initial slope $(b - a/RT) = 0$, reflecting the balance between attractive and repulsive forces. When the compression factor is expanded in terms of reciprocal volume in this way (or in terms of pressure), the resulting series is known as the *virial*

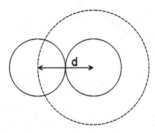

Fig. 9.10. The volume excluded in a binary collision of hard sphere molecules. The volume excluded per molecule, $2\pi d^3/3$, is four times the volume of the molecules.

expansion.

$$\frac{PV_m}{RT} = 1 + \frac{B(T)}{V_m} + \frac{C(T)}{V_m^2} + \dots,$$

where $B(T)$ and $C(T)$ are known as the second and third virial coefficients, respectively. The second virial coefficient, $B(T)$, can be directly related, in a simple manner, to the intermolecular potential energy, $u(r)$, by

$$B(T) = -2\pi N_A \int \left\{ \exp\left[\frac{-u(r)}{kT} \right] - 1 \right\} r^2 dr$$

and is an important source of information about intermolecular forces.

9.6 Critical behaviour

It is found that, below a critical temperature, characteristic of each substance, discontinuities occur in the plot of the pressure of a gas against volume, indicating that the increasing pressure is causing liquefaction. The nature of this critical point was established by Andrews who measured pressure as a function of volume for CO_2 at a series of temperatures (Fig. 9.11). He found that above a critical temperature T_C, no amount of pressure would cause the gas to liquefy and the gas could be taken to high densities, characteristic of the liquid, with no discontinuity. At temperatures below the critical temperature, the gas can only be compressed to a certain pressure (the vapour pressure of the liquid) above which liquefaction will proceed. The vapour pressure of the liquid depends on temperature and has a maximum value, called the

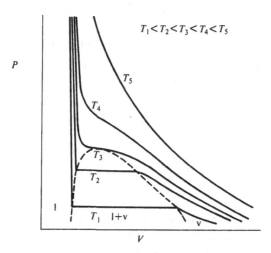

Fig. 9.11. Isotherm for an imperfect gas near to the critical point. T_3 is the critical temperature. The coexistence curve for liquid, l, and vapour, v, is indicated by the dotted line.

critical pressure P_C, at the critical temperature. At the critical point, (T_C, V_C, P_C), both $(\partial P/\partial V)_T$ and $(\partial^2 P/\partial V^2)_T$ are zero. The van der Waals equation gives a simple, but not exact, account of this behaviour and, differentiating the equation with respect to volume and setting both the derivatives to zero, we obtain

$$a = \frac{9RT_C V_C}{8} = 3P_C V_C^2 \quad \text{and} \quad b = \frac{V_C}{3}.$$

9.7 Corresponding states

Using the above relationships between the constants in the van der Waals equation and the critical properties, we can write the equation, for one mole of gas, in the form

$$\left[\frac{P}{P_C} + \frac{3}{(V/V_C)^3}\right]\left[3\left(\frac{V}{V_C}\right) - 1\right] = \frac{RT}{T_C}.$$

The ratios P/P_C, V/V_C and T/T_C are called the *reduced* pressure, volume and temperature. If properties are expressed in terms of these reduced values, then the equation of state of all substances becomes the same (Fig. 9.12). This is a statement of the *principle of corresponding states*. Its generality is not restricted to the predictions of the van der Walls equation and a more general form of the principle can be obtained from a consideration of intermolecular energies. If the intermolecular potential energy can be written in the form

$$u(r) = \varepsilon f\left(\frac{r}{\sigma}\right),$$

where f is a universal function, such as the Lennard–Jones potential function introduced earlier, the properties can be expressed in terms of ε and σ. We define a reduced temperature by $T^* = kT/\varepsilon$ and a reduced volume in terms of $r^* = r/\sigma$, giving $V^* = V/N_A\sigma^3$.

Writing the intermolecular energy as $u^*(r^*) = f(r^*, T^*)$, we can obtain, for the second virial coefficient,

$$B^*(T) = 3\int_0^\infty \left[\frac{1 - u^*}{T^*}\right] r^{*2} dr^* = \frac{B(T)}{\left[\left(\frac{2}{3}\right)\pi N_A\sigma^3\right]}.$$

Thus, if the potential energy function is a universal function when expressed in reduced form in terms of ε and σ, then the second virial coefficient would be a universal function. All substances with potential energy that can be represented by the same universal function will have the same *reduced* second virial coefficient B^*

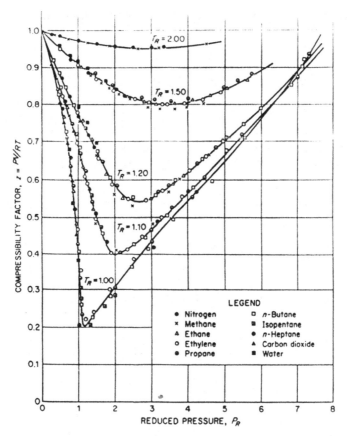

Fig. 9.12. The compression factor as a function of reduced pressure, $P_R = P/P_C$ at a number of reduced temperatures, $T_R = T/T_C$.

at the same reduced temperature T^*. This principle is a reasonable approximation for very simple substances and has greatly facilitated the investigation and representation of the forces between molecules.

9.8 The liquid state

The structure of liquids is more difficult to define than that of solids and gases. The molecules are not fixed in a regular array as in solids, nor are they free to move randomly through space as in gases. One way of representing the structure is in terms of the number of molecular centres per unit volume as a function of distance from the centre of a reference molecule. Clearly, this function starts out as zero, as two molecules cannot overlap, owing to their repulsive forces. At large distances, it must revert to the average number density, N/V, where N is the number of molecules contained in a total volume V. However, at intermediate distances, it will reflect the way the molecules congregate. We define this function, the *pair distribution function*, g(r), by

$$g(r) = \frac{n(r)V}{N^2 4\pi r^2 dr},$$

where $n(r)$ is the number of molecules in the volume $4\pi r^2 dr$ at a distance r from the reference molecule (Fig. 9.13). If the intermolecular potential energy function is $u(r)$ and the molecules interact only in pairs, the total intermolecular energy u_N

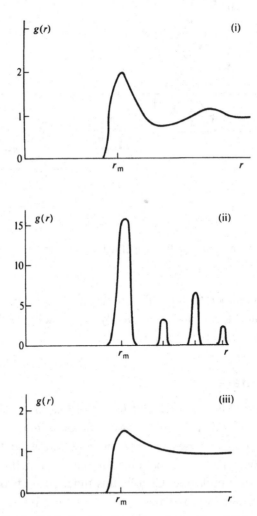

Fig. 9.13. The radial pair distribution functions for monatomic substances: (i) liquid; (ii) solid; (iii) dilute gas.

is given by

$$u_N = \left(\frac{2\pi N^2}{V}\right)\int_0^\infty u(r)g(r)r^2 dr.$$

The calculation of the properties of liquids proved a great challenge even when it was possible to make a reasonable estimate of $u(r)$. The difficulty was that, even with knowledge of $u(r)$, the structure of the liquid, represented by $g(r)$, was still impossible to calculate with sufficient accuracy. However, a successful theory of the liquid state was developed using a perturbation theory of condensed matter not dissimilar to that employed in quantum mechanical calculations (Section 4.8). This theory assumed that, when the force law within the system, that is $u(r)$, is changed, the structure, represented by $g(r)$, will change less rapidly than the energy of the system. Then, if we could determine $g_0(r)$ for an approximate model, we could use the approximate pair distribution function to make a reasonable estimate of the energy and other properties of liquids. This approach was made a practical possibility when the compression factor and the pair distribution functions for hard sphere molecules were determined by computer simulation (both directly and by the validation of theoretical expressions) and could be used as the starting point, $g_0(r)$. We write, for the energy of the liquid, $U_N = U_0 + U_1$. For hard spheres, $U_0 = 0$. The perturbation theory gives for the Helmholtz free energy, A,

$$A_N = A_0 + \left(\frac{2\pi N^2}{V}\right)\int U_1(r)g_0(r)r^2 dr,$$

where A_0 is the Helmholtz free energy for hard sphere molecules. This approach, in a rather more refined form, provided the breakthrough that enabled the properties of liquids to be calculated. If only the first perturbation term is retained, we obtain an equation of state of the form

$$\left(\frac{PV}{RT}\right) = f_1(V) + \frac{f_2(V)}{T},$$

where the first term is simply the compression factor of a hard sphere fluid, $(PV/RT)_0$. It can be seen that the van der Waals equation (Section 9.5) has the same form where $f_1(V) = V/(V-b)$ and $f_2(V) = -a/VR$. This is a significant factor in the success of the van der Waals equation.

9.9 The solid state

Most solids are crystalline and are characterised by the high degree of regularity that arises from the atoms, molecules and ions being located on lattices. Disordered solids can also occur when the disorder of the liquid state is 'frozen in' by rapid cooling to form glasses. Crystals are classified according to the symmetry of their lattice and by the nature of the forces which hold the particles together. There are four types of force that act in solids.

(i) Ionic forces: In the crystals of NaCl, NaOH etc. the ions are bound by strong coulombic forces. Such crystals tend to have high lattice enthalpies, and are often soluble in water.

(ii) Covalent bonds: Crystals such as diamond are held together by covalent linkages acting between specific pairs of atoms. They typically have a coordination number of four or less and tend to be hard, but can fracture along well-defined planes.

(iii) Van der Waals forces: These are dominant in solids such as solid argon and organic solids such as naphthalene. The forces are comparatively weak, the crystals are soft and have low melting points.

(iv) Metallic bonding: This bonding is due to free electrons which can travel through the array of the positively-charged metal atoms. These solids are dense, strong and are good conductors of heat and electricity.

9.10 Crystal structure

Crystals are characterised by their periodic structure. They are made up of repeating patterns of *unit cells*. These are the smallest units which, when repeated, will generate the extended structure of the crystal. A simple example would be a cubic array with atoms at each corner. Each unit cell contains only one atom (each atom on the corner of the cube contributes 1/8). Somewhat surprisingly, there are only 14 possible unit cells that can fit together to produce a lattice. They were first identified by August Bravais in 1849 and are called *Bravais lattices* (Fig. 9.14). The 14 unit cells are comprised of seven classes identified according to the symmetry of their principal axes. Unit cells may contain more than one atom. For example, though the simple cubic lattice unit cell contains only one atom, the body-centred cubic unit cell contains two and the face-centred cubic unit cell four atoms.

Lattices can be regarded as being made up of sets of individual points containing atoms. They can also be regarded as sets of parallel planes, each set separated by the same characteristic distance and each containing atoms located on lattice points. The planes which we observe as crystal faces are those which contain the highest density of lattice points. Understanding the planes within crystals is an important step which helps us determine crystal structures by diffraction methods. The crystal lattice can be defined in terms of three axes, x, y and z and the angles between these axes. Along each of the axes, the unit cell will repeat at a characteristic distance determined by its dimensions. We label these distances a, b and c. To illustrate how

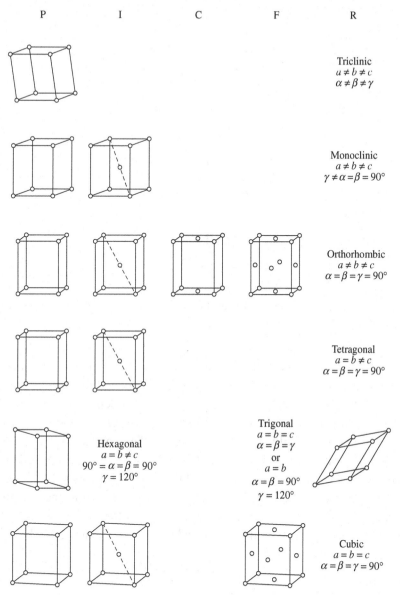

P   I   C   F   R

Triclinic
$a \neq b \neq c$
$\alpha \neq \beta \neq \gamma$

Monoclinic
$a \neq b \neq c$
$\gamma \neq \alpha = \beta = 90°$

Orthorhombic
$a \neq b \neq c$
$\alpha = \beta = \gamma = 90°$

Tetragonal
$a = b \neq c$
$\alpha = \beta = \gamma = 90°$

Hexagonal
$a = b \neq c$
$90° = \alpha = \beta = 90°$
$\gamma = 120°$

Trigonal
$a = b = c$
$\alpha = \beta = \gamma$
or
$a = b$
$\alpha = \beta = 90°$
$\gamma = 120°$

Cubic
$a = b = c$
$\alpha = \beta = \gamma = 90°$

Fig. 9.14. The 14 Bravais lattices. These 14 unit cells generate all the possible three-dimensional crystal lattices. The lattices are organised into five columns. P refers to a primitive unit cell (one lattice point per unit cell), I refers to a body-centred unit cell, C refers to an end-centred unit cell, F refers to a face-centered unit cell, and R refers to a rhombohedral unit cell. (See McQuarrie and Simon, *Physical Chemistry*, University Science Books, Sausalito, p. 1186.)

planes are defined, we will confine ourselves to lattices which have the three axes at right angles (although this is not the case for all the lattices illustrated in Fig. 9.14). We can describe any face on the crystal by the manner in which it intersects the axes. For example, the shaded plane illustrated in Fig. 9.15a intercepts the y and z axes at

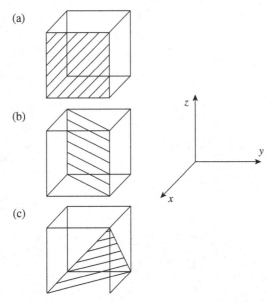

Fig. 9.15. Planes in cubic crystals defined by Miller indices. (a) 100 plane, (b) 110 plane, (c) 111 plane.

infinity and the x axis at the unit cell length, a. In order to define the plane, we take the reciprocals of the intercepts as multiples of the unit cell dimension, in the case we are considering, a. This gives us three integers that define not only the shaded plane but can also be used to define the set of crystal planes equally spaced and parallel to it. For the case illustrated in Fig. 9.15a, the integers are (100). The other two planes illustrated are labelled are (110) and (111). The integers which define the planes are known as the *Miller indices* of the plane and are usually written (hkl). The faces with the smallest Miller indices contain most atoms and are the most important faces of the crystal. The distances between the planes, d, for the cubic lattice we have been investigating, and for which $a = b = c$, are given by

$$\frac{1}{d^2} = \frac{(h^2 + k^2 + l^2)}{a^2}.$$

For example, $(h\,0\,0)$ planes, for which k and l are zero, are separated by a distance of a/h. More generally, for lattices of lower symmetry but for which all the axes are at right angles, the separations of the planes in the crystal are given in terms of Miller indices by

$$\frac{1}{d^2} = \left(\frac{h^2}{a^2} + \frac{k^2}{b^2} + \frac{l^2}{c^2}\right).$$

Because the planes are defined in terms of reciprocal quantities, *those planes with higher Miller indices are closer together.* The separation of planes (222) are exactly half that of the (111) planes.

9.11 X-ray diffraction

In 1912, it was recognised by Max von Laue that X-rays were of an appropriate wavelength to be diffracted by crystals and that the separation between the planes in crystals could be determined from the diffraction patterns that were obtained. Almost immediately, William and Lawrence Bragg, father and son, developed the technique in an important way by employing monochromatic X-rays. Monochromatic X-rays can be produced by the bombardment of elements with high-energy electron beams. Copper is commonly employed as a target and emits X-rays with the principal peak at 154 pm, a very convenient wavelength which is on the same order as the separation of planes within crystals. If we imagine that each plane in the crystal acts as a mirror which reflects X-rays, the patterns of interference that result can be understood (Fig. 9.16). The method of analysis employs what is known as the *Bragg equation.* The rays AD and BC are reflected in the same direction by adjacent planes in the crystal but with a path difference, at plane G, of

$$l = EC + CF = 2EC.$$

$l = 2d \sin \theta$, where θ is the angle the incident beam makes with the planes in the crystal. For the two rays to reinforce, the path difference must be a multiple of the wavelength λ, and

$$n\lambda = 2d \sin \theta.$$

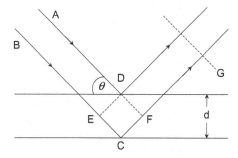

Fig. 9.16. The generation of a diffraction pattern by the reflection of X-rays from planes within a crystal. The path difference, l, between the two rays illustrated is given by $l = EC + CF = 2EC = 2d \sin \theta$.

The first-order reflection with a positive reinforcement of intensity which is observed in the diffraction pattern occurs when $n = 1$. The separation between the various planes in crystals depends on the crystal symmetry. Those planes that contain heavy atoms or ions will reflect X-rays more strongly than those containing only light atoms and ions, since X-rays are scattered by the electrons in the atoms and ions.

The most convenient way of determining the symmetry of crystals and the separation of their lattice planes is by examining the scattering of X-rays from a powdered sample, which will contain a large number of small crystals oriented in all possible directions. Those crystals that are orientated in the correct angles to the beam will produce cones of diffracted radiation at twice the Bragg angles appropriate to the various planes. For each plane, the scattering will be observed at Bragg angles given by

$$\sin^2 \theta = \left(\frac{\lambda^2}{4a^2} \right) (h^2 + k^2 + l^2).$$

For a *simple cubic lattice*, a plot of the intensity of the diffracted X-rays against $\sin^2 \theta$ gives a series of equally spaced lines but, as no three integers can give the result $(h^2 + k^2 + l^2) = 7$, the seventh line is missing. (Also missing would be the 15th, 23rd etc. lines, which also cannot be defined in terms of the sum of the squares of three integers.) (Table 9.2.)

Table 9.2. Diffraction patterns.

$h^2 + k^2 + l^2$	Miller indices	Sum	Simple cubic	Body-centred cubic	Face-centred cubic	CsCl	NaCl	KCl
1	100	1	++	—	—	+	+	—
2	110	2	++	++	—	++	+	—
3	111	3	++	—	++	+	++	—
4	200	4	++	++	++	++	++	++
5	210	3	++	—	—	+	+	—
6	211	4	++	++	—	++	+	—
8	220	4	++	++	++	++	++	++
9	221	5	++	—	—	+	+	—
	300	3						
10	310	4	++	++	—	++	+	—
11	311	5	++	—	++	+	+	—
12	222	6	++	++	++	++	++	++
13	320	5	++	—	—	+	+	—
14	321	6	++	++	—	++	++	—
16	400	4	++	++	++	++	++	++
20	420	6	++	++	++	++	++	++

++ strong
+ moderate
— absent

This information can confirm that the crystal has a simple cubic unit cell and the separation between the lines used to determine its dimensions. The distances between the planes are found to be in the ratios

$$\frac{1}{d_{100}} : \frac{1}{d_{110}} : \frac{1}{d_{111}} = 1 : \sqrt{2} : \sqrt{3}.$$

These ratios are characteristic of the lattice form.

Crystals of different symmetry will give rise to different patterns of diffraction. For example, in a *body-centred cubic lattice*, a plane containing atoms (200) lies midway between the two (100) planes. The intermediate plane causes destructive interference which leads to reflections from the (100) plane being reduced, or completely cancelled out, in the case of the crystal consisting of only one species. A more complete analysis shows that reflections from all planes for which the *sum* $(h + k + l)$ *is odd are missing or weak* (Table 9.2). For a body-centred cubic lattice, we find that the ratios of the spacings between the planes are given by

$$\frac{1}{d_{100}} : \frac{1}{d_{110}} : \frac{1}{d_{111}} = 1 : \frac{1}{\sqrt{2}} : \sqrt{3}.$$

For *face-centred cubic lattices*, similar arguments show that, again, the (100) reflection is weak or entirely absent, but reflections from the (110) plane are also weak or absent, again due to another intermediate plane of atoms giving rise to interference. We find that, for face-centred cubic crystals, strong reflections only occur for planes when h, k and l are *all odd or all even*. For a face-centred cubic lattice, we obtain, for the spacings between the planes,

$$\frac{1}{d_{100}} : \frac{1}{d_{110}} : \frac{1}{d_{111}} = 1 : \sqrt{2} : \frac{\sqrt{3}}{2}$$

When atoms of different types are present, the interference between the planes may not be total, so that lines which are absent for atoms of identical scattering powers now become weaker, but are not totally absent. Thus, for the body-centred cubic lattice of caesium chloride, lines for which the values of $(h + k + l)$ are odd are present, but are weaker. For the face-centred cubic lattice of sodium chloride, all reflections occur, but only those for which h, k and l are either all odd or are all even are of full intensity. The face-centred lattice of potassium chloride provides an interesting case. Because K^+ and Cl^- have the same number of electrons and equal scattering power, KCl appears to be a simple cubic lattice of side $a/2$. The reflections occur in all the planes where the values of h, k and l are twice those of a simple cubic lattice of side a (Table 9.2).

9.12 Molecular structures by diffraction methods

The structure of simple atomic or ionic crystals can usually be established by inspection of the X-ray diffraction patterns and the dimensions of the unit cell easily calculated. However, X-ray diffraction is a much more powerful tool and can be used to determine the molecular structure of large molecules to high precision. In these circumstances, the unit cells can contain not just a few atoms, but very large numbers. The intensity of the radiation scattered from each plane, I_{hkl}, can be related to a *structure factor*, F_{hkl}, such that $I_{hkl} = F_{hkl}^2$. The structure factor depends on the atoms in the unit cell and their geometric arrangement. We can define the structure factor in terms of the *atomic scattering factors*, f_i, and a geometric structure factor, G, which is a function of the coordinates of each atom and which determines the phases of the scattered radiation, $F_{hkl} = \sum f_i G(x_i y_i z_i)$. Each atom will scatter X-rays (at low angles) with a scattering factor f directly proportional to the number of electrons the atom contains (which is equal to the atomic number Z for neutral species). If we know the atomic scattering factors and the positions of the atoms in the unit cell, we can calculate the patterns of intensity formed by the diffracted X-rays.

Unfortunately, the inverse process is much more difficult and a detailed discussion is beyond the scope of this work. Progress is achieved by making use of the fact that any periodic function, such as the electron density in the crystal, can be represented by a series of sine and cosine waves of varying amplitudes and wavelengths. Such a series is termed a *Fourier series*. This enables us, by a Fourier transform, to relate the structure factor to the electron density throughout the unit cell and display the result as a Fourier contour map of the electron density (Fig. 9.17).

A most important problem in determining the structures from X-ray measurements, referred to as the *phase problem*, arises from the fact that the observed intensities are proportional to the square of the structure factor. Thus, we are able to determine the numerical values of the structure factor, but not its sign. To do this, we would have to be able to measure the phase of the scattered radiation. The determination of molecular structures is therefore dependent on additional information. One method of determining the structures involves the presence of a naturally occurring or added heavy atom. This will determine many of the phases required. An alternative method, which has become increasingly useful as computer power has increased, is called the *Patterson synthesis*. In this case, a series involving the intensities rather than the structure factor is summed. The resulting contour map gives not the positions of the atoms, but the distances between the atoms. This leads to $1/2\ N(N - 1)$ peaks, rather than the N peaks of the Fourier map. Taken in conjunction with chemical knowledge, such as typical bond lengths, this may enable the molecular structure to be determined

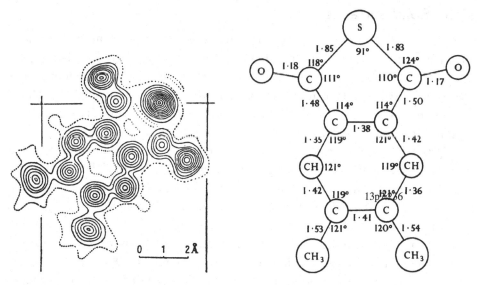

Fig. 9.17. Fourier contour map of the electron density in 4,5-dimethyl phthalicthioanhydride.

unequivocally. In general, a technique that involved much difficult and tedious mathematical manipulation has been liberated and enhanced by the advent of electronic computers. Now, the structures of molecules with many atoms can often be determined directly without resort to additional information. X-ray crystallography has provided us with the structure of many large, biologically-important molecules and has made an enormous contribution to our understanding of biological processes.

The diffraction of both neutrons and electrons has also been used to determine molecular structures. The diffraction of electrons was largely employed in the gas phase to obtain the dimensions of small molecules to high accuracy. Currently, it is used to study solid surfaces and very thin solid-state samples. Neutrons have played an important part in molecular structure determinations and the information obtained from their scattering is often complementary to that obtained from X-rays. Neutron beams are obtained from atomic piles and reflected from a single crystal to provide a beam of a single wavelength. Whereas X-rays are scattered primarily by electrons, neutrons are scattered either by nuclei or by the magnetic moments of atoms. The scattering of neutrons varies relatively little from one nucleus to another and does not fall off with the angle as does the scattering of X-rays. This is particularly helpful in studying compounds containing hydrogen atoms whose scattering is often masked by heavier elements in X-ray diffraction. Neutrons can also provide information about the way in which the magnetic moments of atoms are ordered in ferromagnetic and anti-ferromagnetic substances.

9.13   Solid surfaces

The surfaces of solids play an important part in chemistry. The atoms at the surface have unsatisfied valencies and are thus more chemically active. For this reason, surfaces play an important role by catalysing many of the most important industrial processes. Surfaces are far from uniform and have defects, steps and terraces with corners often originating from dislocations in the crystal below. The growth of crystals is controlled by the surface defects.

Surfaces are rarely in a pure state. Much of the difficulty in studying them experimentally is the necessity for high-vacuum conditions ($\sim 10^{-9}$ Pa) and for high-temperature treatments to drive off adsorbed molecules. Once a clean surface can be established, a number of experimental techniques are available to investigate the structure.

Low-energy electron diffraction (LEED) employs a beam of electrons in the energy range 1 eV to 200 eV with wavelengths in the region 100 pm to 400 pm. At these wavelengths, the predominant diffraction is from the surface and the scattered electrons enable the two-dimensional structure of the surface to be examined. The technique enables the differences between the structure of the surface and that of the bulk solid to be investigated.

A powerful tool for investigating surfaces is scanning tunnelling microscopy. The basis of the procedure rests on the ability of the electronic wave functions to penetrate a potential barrier in a manner that would be forbidden classically. A sharp tip is positioned within a fraction of a nanometre from the surface and a potential difference is applied between the sample and the tip. By measuring the current flow from the tip to the surface, or, alternatively, by measuring the separation of the tip required to maintain a constant current, the character of the surface can be mapped. A resolution of 250 pm can be obtained, showing the structure of surfaces at an atomic level. The technique is extremely flexible and can be employed from ultra-high vacuum to pressures of several atmospheres and from liquid helium temperatures to over 1000 K. Surfaces can also be studied by a wide variety of spectroscopic techniques, including infrared, Raman and photo-electron spectroscopy. Molecular beams have also been used to investigate surfaces and adsorbed species.

The adsorption of gases on solid surfaces may occur in two ways. *Chemisorption* occurs when chemical bonds (usually covalent) are formed, comparable in strength to the ones found in chemical compounds, between the gas molecule and the surface. Alternatively, gas molecules may become attached to the surface by the process of *physisorption*, where physical, van der Waals, forces provide the binding. These forces are generally much weaker and, unlike chemisorption, physisorption is easily reversible, for example when the pressure of the adsorbing gas is reduced.

The most straightforward theory of the adsorption of gases on solids was attributed to Irving Langmuir in 1916. He assumed that all sites on the surface were equivalent and that the gas molecules would be adsorbed until the surface was covered by a molecular monolayer. The rate of adsorption would be proportional to the pressure, which determines the number of collisions gas molecules have with the surface, and the fraction of the surface remaining uncovered. The rate of desorption was assumed to be directly proportional to the fraction of the surface covered by the adsorbed species. At equilibrium, the rate of adsorption and the rate of desorption become equal. If we designate the fraction of the surface covered as θ then, at pressure P,

$$Nk_1 P(1 - \theta) = Nk_2\theta,$$

where k_1 and k_2 are rate constants for adsorption and desorption, respectively, and N is the total number of adsorption sites on the surface. The surface coverage may be expressed as

$$\theta = \frac{k_1 P}{(k_2 + k_1 P)} = \frac{KP}{(1 + KP)}, \quad \text{where } K = \frac{k_1}{k_2}.$$

This equation is known as the *Langmuir isotherm* (Fig. 9.18). At low pressures, θ is proportional to the pressure, since $\theta \approx KP$, whereas, at high pressures, $(1 + KP) \approx KP$, giving $(1 - \theta) \approx 1/KP$ and $\theta \to 1$.

The Langmuir isotherm can also be expressed in terms of the volume of gas adsorbed. It is common to express the fraction of the surface covered, θ, in terms of the volume of gas adsorbed, V, corrected to the standard pressure of one bar. Thus, $\theta = V/V_{max}$, where V_{max} is the maximum volume adsorbed, when it can be assumed

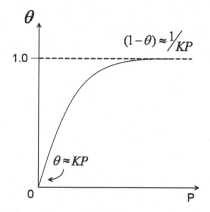

Fig. 9.18. The Langmuir isotherm. θ represents the fraction of the surface covered as a function of the pressure of the gas adsorbed. $\theta = KP/(1 + KP)$.

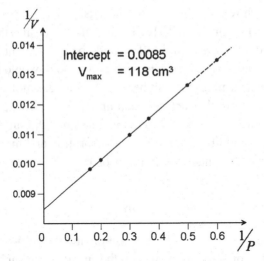

Fig. 9.19. Langmuir plot for nitrogen absorbed on silica at 90 K. The quantity of nitrogen is given in terms of its volume at 298 K and 1 bar pressure.

that the surface is totally covered. Substituting in the original equation, we obtain, after writing $K = k_1/k_2$,

$$\frac{P}{V} = \frac{P}{V_{\text{max}}} + \frac{1}{KV_{\text{max}}}.$$

A plot of $1/V$ versus $1/P$ gives a straight line with the intercept $1/V_{\text{max}}$ (Fig. 9.19). A determination of V_{max} enables us to calculate the maximum number of molecules adsorbed,

$$n_{\text{m}} = \frac{N_A P V_{\text{max}}}{RT},$$

where P is the standard pressure (1 bar). From a knowledge of the collision diameter of the molecules, we can estimate the surface area of the sample, $A = n_{\text{m}}\sigma^2$.

When chemical reactions take place on surfaces, it is often assumed that the rate of a reaction such as

$$A + B \rightarrow \text{products}$$

can be expressed in terms of the fraction of the surface covered by each of the reactants, giving a rate proportional to the product $\theta_A \theta_B$. This approach will be developed further in Chapter 11.

In many cases of adsorption, especially physisorption, the data are not fitted accurately by the Langmuir isotherm. Nevertheless, it provides the most widely used starting point for the interpretation of the interaction of gases with solid surfaces. An alternative isotherm, which often provides a superior description for physisorption, is that devised by Brunauer, Emmett and Teller, known as the BET isotherm. It assumes that, once the surface is covered with a monolayer, further layers may continue to be deposited forming an adsorbate that has much in common with the liquid state of the adsorbing substance. In this model, there is no limit to the amount that can be adsorbed. The volume adsorbed may be expressed

$$\frac{V}{V_{\mathrm{mon}}} = \frac{cf}{(1-f)[1+(c-1)f]},$$

where $f = P/P^*$ and P^* is the vapour pressure of the adsorbate at the temperature of the experiment, V_{mon} is the volume required to give monolayer coverage of the surface and $c = \exp[(\Delta_{\mathrm{ad}}H - \Delta_{\mathrm{vap}}H)/RT]$. $\Delta_{\mathrm{ad}}H$ is the enthalpy of desorption when there is only a monolayer on the surface. In the case of a single monolayer, the BET isotherm becomes equivalent to the Langmuir isotherm. In general, the shape of the isotherm depends critically on the value of the parameter c.

9.14 Summary of key principles

The state of matter in which a substance exists at a particular temperature and pressure is that of the lowest free energy. Since $G = H - TS$ and $(\partial G/\partial T)_P = -S$, gases, which possess the highest entropy, are the stable form at high temperatures; solids with a low entropy and high (negative) enthalpy are stable at low temperatures. Discontinuous, isothermal phase transitions, such as melting and vaporization, occur when the free energy of two phases becomes equal. Then, $dG = 0$ and $\Delta S = \Delta H/T$. The boundaries between the phases can be represented by *phase diagrams* on which a single point defines the specific conditions at which all three phases can be in equilibrium (*the triple point*). Above a well-defined *critical temperature*, no discontinuous change between liquids and gases is observed.

The behaviour of matter is determined by intermolecular forces and ionic forces. At short range, the overlap of the electron orbitals leads to steep repulsion. At long range, *electrostatic, induction and dispersion forces* lead to an attractive energy. It is not possible to calculate these forces for other than the simplest of atoms and molecules.

The structure of liquids can be described in terms of a *pair distribution function*, $g(r)$, which describes the probability of finding the centre of one molecule at a distance r from the centre of another. The properties of liquids can be calculated by assuming that the attractive forces can be treated as a perturbation and assuming that the repulsive forces alone determine the structure of the liquid. This assumption leads to equations of the form $PV/nRT = f_1(V) + f_2(V)/T$. An early equation of state of this type was that of van der Waals, who proposed that $f_1(V) = V/(V-b)$ and $f_2(V) = -a/VR$. This equation gave a good account of the behaviour of liquids including the existence of the critical temperature. Computers now allow for more accurate evaluation of these functions.

Solids are normally characterised by regular arrays which can be described by the arrangements of the planes on which the atoms, ions or molecules lie. The planes are identified by the reciprocal of their intercepts on the x, y and z axes (*Miller indices*). These arrangements and the distances between the planes can be determined experimentally from the interference patterns generated by the scattering of X-rays, which have a wavelength on the same order as the separation of crystal planes. The study of X-ray diffraction has contributed greatly to the determination of the structures of large molecules of biological importance. Solid surfaces have played an important part in chemistry because of their high reactivity which enables them to catalyse many processes of industrial importance.

Problems

(1) The vapour pressure of a liquid is 0.0526 bar at 280 K and 0.132 bar at 298 K. Calculate the enthalpy of vaporization of the liquid.

(2) At 273 K the density of water is $1.00 \, \text{g cm}^{-3}$ and that of ice $0.917 \, \text{g cm}^{-3}$. Estimate the change in the melting point of ice when a pressure of 200 bar is applied. (For water, $\Delta_{\text{fus}} H = 6.0 \, \text{kJ mol}^{-1}$.)

(3) The diffraction of X-rays of wavelength 0.1537 nm from the (100) planes of a simple cubic crystal gave a first-order reflection at $\theta = 11.45°$. Calculate the separation between the planes.

(4) The first-order reflections from the (100), (110), (111) planes of a crystal were observed at 5.9°, 8.4°, and 5.2°. Determine the symmetry of the unit cell.

(5) The volume of a gas (measured at 298 K and 1 bar) adsorbed on a sample of charcoal was:

P/atm	0.0132	0.0395	0.132
V/cm^3	45.0	60.2	68.4

Show that the data are consistent with the Langmuir isotherm. Estimate the surface area of the charcoal, assuming that the diameter of the gas molecules is 500 pm.

10

Mixtures and Solutions

Many chemical experiments are carried out in solution and it is therefore very important that we are able to understand the behaviour of substances in solution. The simplest model of a mixture is the *ideal solution*. Such a solution is formed when the components have molecular interactions which are virtually identical. It is a severe approximation, but it provides the platform from which all real solutions are approached.

10.1 The ideal solution

Consider the process where two very similar, but distinct, substances are mixed. The mixing leads to an increase in entropy which arises because, whereas the interchange of molecules in a pure compound does not produce a different microstate, the exchange of molecules in a mixture does. If we assume the components are sufficiently similar to mix randomly, then

$$W_{mix} = \frac{(n_1 + n_2)!}{n_1! n_2!}$$

(see Section 7.4). However, since $S = k \ln W$ and Stirling's approximation, when N is large, gives $\ln N! = N \ln N - N$, we obtain, for the entropy arising from the mixing process,

$$\Delta_{mix} S = k[(n_1 + n_2) \ln(n_1 + n_2) - (n_1 + n_2) - n_1 \ln n_1 - n_2 \ln n_2 + n_1 + n_2].$$

This can be rearranged to give,

$$\Delta_{mix} S = k \left\{ (n_1 + n_2) \left[-\left(\frac{n_1}{(n_1 + n_2)} \right) \ln \left(\frac{n_1}{(n_1 + n_2)} \right) \right. \right.$$
$$\left. \left. - \left(\frac{n_2}{(n_1 + n_2)} \right) \ln \left(\frac{n_2}{(n_1 + n_2)} \right) \right] \right\}.$$

231

The mole fraction of component one is $x_1 = n_1/(n_1 + n_2)$, that of component two is $x_2 = n_2/(n_1 + n_2)$ and $(n_1 + n_2) = N$, so we obtain

$$\Delta_{\text{mix}} S = -Nk(x_1 \ln x_1 + x_2 \ln x_2)$$

and, for one mole, when $N = N_A$,

$$\Delta_{\text{mix}} S = -R(x_1 \ln x_1 + x_2 \ln x_2).$$

$x_i < 1$ and so $\Delta_{\text{mix}} S > 1$ and the entropy change on mixing is positive.

We assume that, since the molecules making up the mixture are very similar,

$$\Delta_{\text{mix}} H = 0.$$

And, since $\Delta_{\text{mix}} G = \Delta_{\text{mix}} H - T\Delta_{\text{mix}} S$, we obtain, for an ideal solution,

$$\Delta_{\text{mix}} G = RT(x_1 \ln x_1 + x_2 \ln x_2).$$

$\Delta_{\text{mix}} G < 1$ and mixing occurs with a negative free energy change and, therefore, is a spontaneous process. Since $\Delta V = (\partial \Delta G/\partial P)_T$, we find, for an ideal solution, that $\Delta_{\text{mix}} V = 0$.

The change in the chemical potential of each component due to the process of mixing is given by

$$\Delta \mu_i = \left(\frac{\partial \Delta G}{\partial n_i} \right)_{T,P,n_j} = RT \ln x_i$$

and, for each component (assuming the ideal solution is obtained by mixing liquids),

$$\mu_i(\text{soln}) = \mu_i^*(\text{l}) + RT \ln x_i,$$

where μ_i^* is the *chemical potential of pure liquid i* in equilibrium with its vapour. This equation, if followed by all components, provides another very useful definition of ideal behaviour.

The original definition of ideal solutions was based on the behaviour of the vapour pressure of the components. For component i, we can write, for the gas phase in equilibrium with the solution,

$$\mu_i(g) = \mu_i^0(g) + RT \ln \left(\frac{P_i}{\text{bar}} \right),$$

where $\mu_i^0(g)$ is the chemical potential of the gas at 1 bar and P_i is the vapour pressure of component i in the solution. At equilibrium, the chemical potential of component

i in the gaseous phase must be equal to the chemical potential of i in the solution. Thus,

$$\mu_i(\text{soln}) = \mu_i^*(\text{l}) + RT \ln x_i = \mu_i^0(\text{g}) + RT \ln \left(\frac{P_i}{\text{bar}} \right).$$

The chemical potential of i in its pure liquid state is equal to its chemical potential in the gaseous phase at its vapour pressure, P_i^*:

$$\mu_i^*(\text{l}) = \mu_i^0(\text{g}) + RT \ln \left(\frac{P_i^*}{\text{bar}} \right).$$

Substituting for $\mu_i^*(\text{l})$, we obtain

$$\mu_i^0(\text{g}) + RT \ln \left(\frac{P_i^*}{\text{bar}} \right) + RT \ln x_i = \mu_i^0(\text{g}) + RT \ln \left(\frac{P_i}{\text{bar}} \right)$$

and

$$RT \ln \left(\frac{P_i}{\text{bar}} \right) = RT \ln \left(\frac{P_i^*}{\text{bar}} \right) + RT \ln x_i.$$

Thus, $P_i = x_i P_i^*$, giving a linear relationship for the total vapour pressure of the solution,

$$P = x_1 P_1^* + x_2 P_2^*,$$

which is known as *Raoult's law*. Adherence to Raoult's law was the traditional definition of an ideal solution. As a definition, it is consistent with those expressed in terms of chemical potential given above.

Under normal pressure conditions, the pressure variation of the chemical potential of the liquid is very small and we can often replace $\mu_i^*(\text{l})$, which is defined at the vapour pressure of the liquid, by $\mu_i^0(\text{l})$, the chemical potential at 1 bar pressure, to obtain

$$\mu_i^*(\text{soln}) = \mu_i^0(\text{l}) + RT \ln x_i.$$

Solutions following this equation over all of the composition range are called *truly ideal solutions*. It is a necessary condition for such solutions that the molecules making up the solution are very similar. This is a condition met by very few solutions.

10.2 Truly ideal solutions

If the components of a solution follow Raoult's law over the whole range of composition, the total vapour pressure is a linear function of the mole fractions

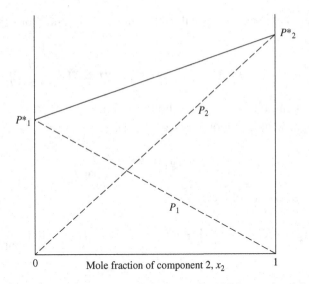

Fig. 10.1. Pressure of the vapour in equilibrium with a liquid mixture that follows Raoult's law. Broken lines are the partial vapour pressures of the components.

(Fig. 10.1). However, the composition of the vapour phase does not follow the same linear behaviour. Consider the composition of the vapour in terms of the mole fraction of component B, the more volatile component, $x_B(g)$.

$$x_B(g) = \frac{P_B}{P} = x_B(l)\frac{P_B^*}{P}$$

and

$$x_A(g) = \frac{P_A}{P} = x_A(l)\frac{P_A^*}{P},$$

where P_A^* and P_B^* are the vapour pressures of the pure liquids and P is the total pressure, $P = P_A + P_B$. Thus,

$$\frac{x_B(g)}{x_A(g)} = \frac{x_B(l)P_B^*}{x_A(l)P_A^*}$$

and the vapour is richer in the more volatile component B than the liquid with which it is in equilibrium. It is this behaviour that allows compounds to be separated by distillation. Consider the process of boiling a liquid mixture with the composition of component B, $x_B(l)$, as illustrated in Fig. 10.2. The vapour in equilibrium with this liquid will have composition $x_B(g)$. If this vapour is condensed, we would obtain a liquid significantly richer in component B. A single step such as this is described as a 'theoretical plate'. Practical distillation columns carry out the equivalent of many such steps and their efficiency is judged by how many theoretical plates they provide.

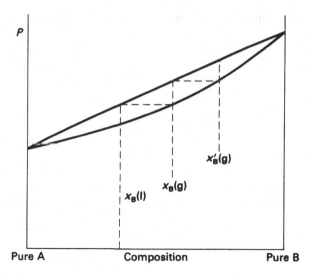

Fig. 10.2. A mixture of two liquids that follows Raoult's law. The total vapour pressure of the mixture is plotted as a function of the composition of the liquid (upper line) and of the vapour (lower line). The upper line is sometimes referred to as the *bubble-point line* and the lower as the *dew-point line*.

10.3 Ideal solutions of solids in liquids

Some solids can form ideal solutions in suitable liquid solvents. To understand the process of dissolution, we can imagine it taking place in two stages.

(i) The first step is melting the solid to produce a liquid (since the temperature will be below the normal melting point of the solid, this is a hypothetical process). The enthalpy change associated with this process is the enthalpy change on fusion, $\Delta_{fus} H$.

(ii) The second step is mixing the melted solid with the solvent. If this forms an ideal solution, $\Delta H = 0$ for this stage.

For a saturated solution, the chemical potential of the solute in the solution will be equal to that of the solid solute, so that

$$\mu_2^0(s) = \mu_2^0(l) + RT \ln x_2$$

and, since

$$\Delta_{fus} \mu^0 = \mu_2^0(l) - \mu_2^0(s),$$

$$\ln x_2 = -\frac{\Delta_{fus} \mu^0}{RT}.$$

Differentiating using the Gibbs–Helmholtz equation (Section 8.3), we obtain

$$\left(\frac{\partial \ln x_2}{\partial T}\right)_P = - \left[\frac{\partial \left(\Delta_{\text{fus}} \mu^0 / RT\right)}{\partial T}\right]_P = - \left[\frac{\partial \left(\Delta_{\text{fus}} G^0 / RT\right)}{\partial T}\right]_P$$

and

$$\left(\frac{\partial \ln x_2}{\partial T}\right)_P = \frac{\Delta_{\text{fus}} H^0}{RT^2}.$$

At the melting point, when the solid goes to its natural liquid state, the two liquids will mix completely (and are said to be completely miscible). Thus, $x_2 = 1$ when $T = T_{\text{fus}}$. We can integrate the equation given above (at constant pressure) to obtain

$$\ln x_2 = \left(\frac{\Delta_{\text{fus}} H^0}{R}\right)\left[\frac{1}{T_{\text{fus}}} - \frac{1}{T}\right].$$

When the solute is very similar to the solvent, the ideal solubility equation is followed fairly accurately. However, most solids differ sufficiently from the solvent to make this approximation of limited applicability. Figure 10.3 shows the results for the solubility of naphthalene in benzene and naphthalene in cyclohexane as a function of temperature. Naphthalene, as an aromatic molecule, is sufficiently chemically

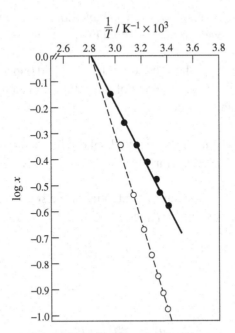

Fig. 10.3. The logarithm of the solubility of solid naphthalene (expressed as a mole fraction) as a function of temperature. • naphthalene in benzene, ○ naphthalene in cyclohexane. The solid line is calculated using the ideal solubility equation.

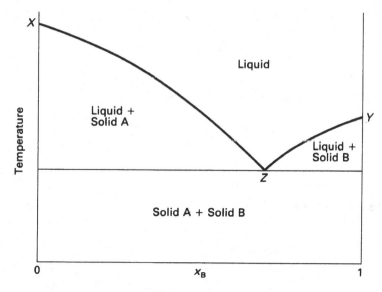

Fig. 10.4. A system showing a eutectic point.

similar to benzene to form an almost ideal solution. With cyclohexane, however, there is less similarity and less naphthalene dissolves than would be predicted by ideal behaviour.

Figure 10.4 illustrates the mutual solubility of two substances A and B. The line XZ represents the solubility of B in A. However, it can also be regarded as defining the freezing point of A, which is reduced by the presence of component B. Thus, if the solid has the composition of the solvent, the substance in greater quantity, we call it a freezing point curve, such as when ice freezes out of a salt solution. If the solid separating out is the solute, we call the line the solubility curve, such as would be observed if, for example, iodine separated out from a solution in benzene. Where the two curves meet, we have what is called a *eutectic point*, the lowest melting point of the mixture. On cooling to this point, the temperature will remain constant until all the liquid has frozen to produce a mixture of solids A and B. The position of the eutectic point can be estimated by applying the ideal solubility equation to both components.

10.4 Ideal dilute solutions

Truly ideal behaviour is very uncommon. However, most solutions show some of the characteristics of ideal behaviour when dilute. In a very dilute solution, both the solvent and solute molecules are surrounded by the solvent, so that the solvent molecules tend, at infinite dilution, to the pure solvent state, but the solute tends to a hypothetical fluid in which every solute molecule is surrounded by a solvent molecule and all the interactions are of the solvent-solute type (Fig. 10.5). The

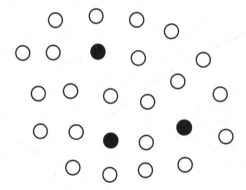

Fig. 10.5. Molecules of solvent, o, and solute, •, in a dilute solution (schematic).

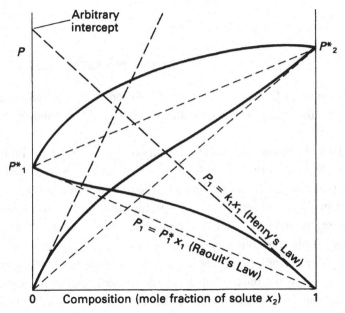

Fig. 10.6. The vapour pressure of the components and the total vapour pressure of a liquid mixture that deviates positively from Raoult's law (schematic). Water-ethanol mixtures deviate from Raoult's law in this manner.

solvent vapour pressure will be directly proportional to its mole fraction, $P_1 = x_1 P_1^*$, where P_1^* is the vapour pressure of the pure liquid solvent (Fig. 10.6). On the other hand, for the solute, the vapour pressure is again proportional to its mole fraction, but with an arbitrary intercept which represents a hypothetical liquid in which all interactions are of the solvent-solute type, and

$$P_2 = x_2 P_2^{\ominus}.$$

$P_2^{\ominus}$ represents the vapour pressure of this hypothetical fluid. Under these circumstances, *the solvent is said to follow Raoult's law and the solute, Henry's law.*

This behaviour is helpful in the understanding of *colligative properties*, properties that depend only on the number of molecules in a solution rather than also on their nature. These properties played an important part in the history of chemistry, providing a method to determine the molecular masses and the nature of molecular species in solution.

One example is the lowering of the vapour pressure of a solvent by a non-volatile solute in a dilute solution. If the solvent follows Raoult's law, then $P_1 = x_1 P_1^*$ and $\mu_1 = \mu_1^* + RT \ln x_1$. If the solute is non-volatile, it does not contribute to the total pressure and enters into this equation only in so far as $x_1 = (1 - x_2)$. We can determine x_2 from knowledge of the masses of the solvent and solute present in the solution, w_1 and w_2, and their molecular masses, M_1 and M_2.

$$x_2 = \frac{n_2}{(n_1 + n_2)} = \frac{w_2/M_2}{(w_1/M_1 + w_2/M_2)}.$$

If w_1, w_2 and M_1 are known, we can determine the molecular mass of the solute in solution by measuring the lowering of the vapour pressure that occurs when it is added to the solvent.

As the presence of the solute lowers the chemical potential of the solvent and consequently its vapour pressure, the boiling point of the solution, the temperature at which the vapour pressure is equal to one bar, is higher than that of the pure solvent (Fig. 10.7). If the solvent in the solution is in equilibrium with the gaseous phase at one bar pressure, we have

$$\mu_1(\text{soln}) = \mu_1^0(l) + RT \ln x_1 = \mu_1^0(g)$$

and

$$RT \ln x_1 = \mu_1^0(g) - \mu_1^0(l) = \Delta_{\text{vap}} G^0.$$

Differentiating this expression using the Gibbs–Helmholtz equation (Section 8.3), we obtain $(\partial \ln x_1/\partial T) = -\Delta_{\text{vap}} H^0/RT^2$ and, integrating from the pure solvent, for which $x_1 = 1$ and $T = T_{\text{vap}}$, to the solution, for which the composition is x_1 and the boiling temperature is T, we obtain

$$\ln x_1 = -\frac{\Delta_{\text{vap}} H^0}{R} \left[\frac{1}{T_{\text{vap}}} - \frac{1}{T} \right].$$

In a very dilute solution, $\ln x_1 = \ln(1 - x_2) \approx -x_2$, and, if the elevation of the boiling point is small, it is usual to approximate $[1/T_{\text{vap}} - 1/T]$ by $\Delta T/T_{\text{vap}}^2$,

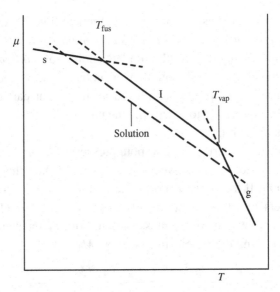

Fig. 10.7. Effect of added solute on the chemical potential of a liquid solvent as a function of temperature. The freezing point is depressed and the boiling point is elevated.

giving

$$\Delta T = \frac{x_2 R T_{\mathrm{vap}}^2}{\Delta_{\mathrm{vap}} H^0}.$$

If the solute does not form a solid solution with the solvent when it freezes out, then the freezing point of the solution will be less than that of the solvent. A very similar reasoning to that given above for the elevation of boiling point gives an expression for the depression of freezing point,

$$\Delta T = \frac{x_2 R T_{\mathrm{fus}}^2}{\Delta_{\mathrm{fus}} H^0}.$$

Another example of a colligative property is the *osmotic pressure*. If a solution is separated from the pure solvent by a membrane which is only permeable to the solvent molecules, then there is a tendency for the solvent molecules to diffuse into the solution. This is because the chemical potential of the solvent in the solution will be reduced by $RT \ln x_1$ due to the presence of solute molecules. This can be counteracted and equilibrium restored if a pressure is applied to the solution to restore the chemical potential of the solvent in solution to that of the pure solvent (Fig. 10.8). The change in the chemical potential of the solvent as a function of pressure is given by

$$\left(\frac{\partial \mu_1}{\partial P} \right)_T = V$$

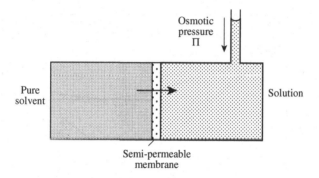

Fig. 10.8. Schematic representation of an osmotic pressure apparatus.

(Section 8.11). The *osmotic pressure*, Π, required to compensate the reduction in the chemical potential of the solvent due to the solute, $RT\ln x_1$, is

$$RT\ln x_1 = \Pi \left(\frac{\partial \mu_1}{\partial P}\right) = \Pi V_1.$$

Similar approximations to those used for the other colligative properties lead to the result for dilute solutions where m is molarity,

$$\Pi = \frac{RT x_2}{V_1} = m \times 10^3 RT.$$

The osmotic pressure is proportional to the concentration and can be used in a similar way to the other colligative properties to determine molecular masses of solids in solution. Osmotic pressure plays an important role in controlling concentrations in biological systems.

Colligative properties contributed greatly to our understanding of the behaviour of substances in solution, particularly when compounds dissociated or associated in solution, as this was reflected in the apparent molecular mass differing from the molecular mass of the monomer. Thus, it was established that sodium chloride in solution is essentially fully dissociated into ions.

10.5 Non-ideal solutions

For most real solutions, Raoult's law does not hold for the solvent and Henry's law for the solute even at relatively low concentrations (Fig. 10.9), and it is convenient to introduce the concept of *activity*, a_i, defined by

$$\mu_i = \mu_i^0 + RT\ln a_i.$$

a_i is thus an *effective concentration* measured relative to that of the standard state. For non-electrolyte solutions, the standard state usually adopted for the solvent is the

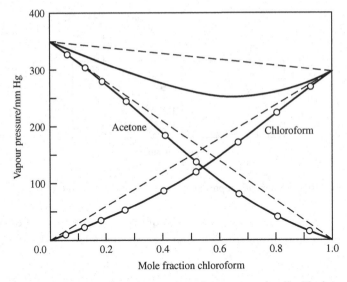

Fig. 10.9. Total vapour pressure and vapour pressures of components in a liquid mixture that deviates negatively from Raoult's law.

pure liquid state, but many other standard states can be used that are more appropriate to a particular type of mixture. The extent to which the effective concentrations are different from the true concentration, x_i, is measured by the *activity coefficient*, γ_i:

$$\gamma_i = \frac{a_i}{x_i} = \frac{\text{effective concentration}}{\text{true concentration}}.$$

The definition of activity enables us to adapt the equations we have developed for ideal solutions to real solutions by replacing x_i by a_i, so that, whereas

$$\text{for ideal solutions, } x_i = \frac{P_i}{P_i^0},$$

$$\text{for real solutions, } a_i = \frac{P_i}{P_i^\ominus},$$

where $\ominus$ represents an appropriate standard state. For most non-ionic liquids that do not hydrogen bond or have strong specific interactions, the deviations from ideality are positive — that is, the vapour pressures are greater than expected. For such solutions, $a_i > x_i$ and $\gamma_i > 1$. The components are less 'happy' in the mixture than we could expect and show a greater tendency to escape to the gaseous phase. In extreme cases, the strong positive deviations from ideal behaviour can cause the two components to separate into two phases.

Solutions of solids in liquids commonly show positive deviations from ideality. In the case of solid naphthalene dissolving in benzene and hexane, we can calculate the activity coefficient of naphthalene in hexane from its solubility. In solutions in both benzene and hexane, the chemical potential of the naphthalene must be that of solid naphthalene and the activity of naphthalene, a_2, must be the same in both solutions, i.e.

$$a_2(\text{benzene}) = a_2(\text{hexane}) \quad \text{and} \quad x_2\gamma_2(\text{benzene}) = x_2\gamma_2(\text{hexane}).$$

In benzene $x_2 = 0.30$, the solution is ideal and the activity coefficient is unity. For hexane $x_2 = 0.12$ so we can estimate the activity coefficient in hexane by (see Section 10.3)

$$\gamma_2 = \frac{x_2(\text{benzene})}{x_2(\text{hexane})} = \frac{0.30}{0.12} = 2.5.$$

10.6   Molecular basis of ideality

It is of interest to consider further what the molecular conditions that lead to ideal solutions are. Let there be a total of N molecules (N_A of species A and N_B of species B), distributed on a lattice, each with z nearest neighbours (Fig. 10.10). (We use a lattice model to simplify visualising the distribution. A lattice is not essential to the argument.) We assume that the molecules are randomly distributed and that they interact only with nearest neighbours. The entropy of mixing is then given by

$$\Delta_{\text{mix}} S = R(x_A \ln x_A + x_B \ln x_B).$$

The energy of the mixture is given by

$$U_{\text{mix}} = \frac{1}{2}z(N_A + N_B)\left[\varepsilon_{AA}x_A^2 + \varepsilon_{BB}x_B^2 + 2\varepsilon_{AB}x_A x_B\right],$$

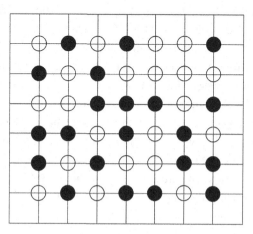

Fig. 10.10.   Two components of a mixture randomly distributed on a lattice. (In this two-dimensional array, each has four nearest neighbours.)

and that of the unmixed components by

$$U_A + U_B = \frac{1}{2}z(N_A + N_B)[x_A \varepsilon_{AA} + x_B \varepsilon_{BB}].$$

The additional energy of the solution, on mixing, is therefore

$$\Delta_{\text{mix}} U = \frac{1}{2}z N[\varepsilon_{AA} x_A^2 + \varepsilon_{BB} x_B^2 + 2\varepsilon_{AB} x_A x_B - x_A \varepsilon_{AA} - x_B \varepsilon_{BB}],$$

where $N = (N_A + N_B)$. Since $(x_A + x_B) = 1$

$$\Delta_{\text{mix}} U = \frac{1}{2}z N[-x_A \varepsilon_{AA}(1 - x_A) - x_B \varepsilon_{BB}(1 - x_B) + 2\varepsilon_{AB} x_A x_B]$$

and

$$\Delta_{\text{mix}} U = \frac{1}{2}z N x_A x_B[2\varepsilon_{AB} - \varepsilon_{AA} - \varepsilon_{BB}].$$

We define

$$\delta = \frac{1}{2}z N[2\varepsilon_{AB} - \varepsilon_{AA} - \varepsilon_{BB}],$$

so that

$$\Delta_{\text{mix}} U = x_A x_B \delta.$$

If the energy of interaction of the unlike molecules is the arithmetic mean of the like-like interactions, so that $2\varepsilon_{AB} = \varepsilon_{AA} + \varepsilon_{BB}$ and $\delta = 0$, we obtain, for the energy of mixing, $\Delta_{\text{mix}} U = 0$. In our simple model, the condition for ideality in a mixture is that the molecules are randomly distributed and that the energy of the unlike interaction is the arithmetic mean of the like-like interactions.

In non-ideal solutions described by this model, the energy of mixing is dependent on the difference in the energy of interaction between the unlike molecules from the arithmetic mean of the like-like interactions. Solutions in which there is a nonzero enthalpy change on mixing and in which the molecules are mixed randomly are called *regular solutions*. Though the enthalpy of mixing is nonzero, it is assumed to be insufficient to change the distribution of the molecules and hence the entropy of mixing remains that of an ideal solution with random mixing. (However, in practice, the entropy of mixing of real solutions is often far from ideal.) If $\Delta_{\text{mix}} V = 0$, we have $\Delta_{\text{mix}} U = \Delta_{\text{mix}} H$ and

$$\Delta_{\text{mix}} G = RT(x_A \ln x_A + x_B \ln x_B) + \delta x_A x_B.$$

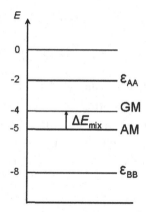

Fig. 10.11. The energies of interaction in a mixture. $\varepsilon_{AA} = -2$, $\varepsilon_{BB} = -8$. The arithmetic mean of the energies of interaction of the two components (AM) corresponds to ideal mixing with a zero energy change on mixing. The geometric mean (GM) is a more realistic approximation leading to a positive energy of mixing given by the difference between the geometric mean energy and the arithmetic mean energy.

It is found that a better estimate of the energy of interaction of the unlike molecules can be obtained by taking a geometric mean of the like interactions, rather than assuming an arithmetic mean, so that

$$\varepsilon_{AB} = (\varepsilon_{AA}\varepsilon_{BB})^{1/2}, \quad \delta = \tfrac{1}{2}zN[2(\varepsilon_{AA}\varepsilon_{BB})^{1/2} - \varepsilon_{AA} - \varepsilon_{BB}]$$

and $\delta = -\tfrac{1}{2}zN[(\varepsilon_{AA})^{1/2} - (\varepsilon_{BB})^{1/2}]^2$. The geometric mean is less than the arithmetic mean and the energy of the mixture is less negative than the energies of the pure components, so that the energy of mixing is positive (Fig. 10.11). Sometimes, regular solutions are defined as solutions in which the unlike interactions are described by the geometric mean rule — a rather narrower definition than that given above. The concept of regular solutions has been employed in a number of ways to enable semi-empirical estimates of the properties of mixtures to be made, but the results are rarely sufficiently accurate to make its results of practical value.

10.7 Ions in solution

An important part of physical chemistry has been the study of ions in solution. Because of the high dielectric constant of water, the attraction between ions is greatly reduced and this allows ionic compounds to dissociate into their constituent ions. Other solvents, such as organic compounds, do not have such a high dielectric constant and do not facilitate this dissociation.

The chemical potential of ionic substances is usually defined relative to a standard state of unit molality (moles per kilogram of solvent). To simplify the notation, we will define m as the dimensionless ratio molality/mol kg^{-1} and write

$$\mu_i = \mu_i^{\circ} + RT \ln m_i \gamma_i.$$

(A word of warning — the standard state is defined, not only by $m_i = 1$, but also requires that $\gamma_i = 1$, leading to an entirely hypothetical state which is equivalent to extrapolating the behaviour of the solution at infinite dilution to one of unit molality.) When we apply this to an electrolyte such as NaCl, which is fully dissociated in solution, we can only determine the overall activity, a_{NaCl}. The activities of the individual ions cannot be measured independently. However, we can express the activity a_{NaCl} in terms of the product of the individual activities a_{Na^+} and a_{Cl^-}:

$$\mu_{total} = \mu_{Na^+}^{\circ} + \mu_{Cl^-}^{\circ} + RT \ln a_{Na^+} + RT \ln a_{Cl^-}$$

If we define the overall activity of the solution by

$$\mu_{total} = \mu_{Na^+}^{\circ} + \mu_{Cl^-}^{\circ} + RT \ln a_{NaCl}$$

then, by comparing the equations, we obtain

$$a_{NaCl} = a_{Na^+} \times a_{Cl^-} = \gamma_{Na^+} m_{Na^+} \times \gamma_{Cl^-} m_{Cl^-}.$$

Since $m_{Na^+} = m_{Cl^-} = m$, the molality of the solution, we obtain $a_{NaCl} = (\gamma_{\pm} m)^2$, where $\gamma^{\pm}$ is termed the mean ionic activity coefficient and is defined by $\gamma_{\pm} = (\gamma_{Na^+} \gamma_{Cl^-})^{1/2}$.

For more complex electrolytes, $M_p A_q = pM^+ + qA^-$, the mean ionic activity coefficient is defined by

$$\gamma_{\pm}^{(p+q)} = (\gamma_{M^+})^p (\gamma_{A^-})^q.$$

10.8 Debye–Hückel theory

What sort of values might we expect for the activities of ions in solution? The commonsense view might be that, as the solution becomes 'cluttered' with other ions, the environment will become less attractive for the ions and the activity coefficients will be positive. This is the exact opposite of what we observe, at least in dilute solutions.

The theory which enables us to quantify this effect was established by Peter Debye and Erich Hückel in 1923. We will content ourselves with setting out some of the main features of their theory. They assumed that the properties of ionic solutions could be interpreted in terms of the electrostatic interactions of ions represented as point charges. Ions of opposite charge experience an attractive force and a negative energy of interaction which leads to a larger Boltzmann factor. This favours unlike

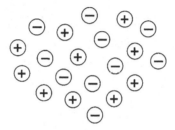

Fig. 10.12. Ions in solution. Illustrating the formation of an ion atmosphere in which each ion is surrounded predominantly by ions of the opposite sign.

ions lying close together so that *each ion is surrounded by an ionic atmosphere in which unlike ions predominate* (Fig. 10.12). Expressing this quantitatively, $n_i = n_i^0$ exp $(-z_i e\Phi/kT)$, where n_i is the local concentration of species i and n_i^0 the bulk concentration, $z_i e$ is the charge on i and Φ is the electrostatic potential. The electrical charge density, ρ, is then given by (summing over all the different ions present)

$$\rho = \sum n_i z_i e = \sum^i n_i^0 z_i e \exp(-z_i e\Phi/kT).$$

If we assume that $z_i e\Phi \ll kT$, we can expand the exponential term by using $e^{-x} = 1 - x + x^2/2! + \cdots$ to obtain

$$\rho = \sum n_i^0 z_i e - \sum^i n_i^0 z_i e (z_i e\Phi/kT) + \cdots$$

The first term, involving the sum of the charges on the ions, $\sum z_i$, is zero because of the overall neutrality of the solution, and we obtain the first nonzero term contributing to charge density,

$$\rho = -\sum n_i^0 z_i^2 e^2 \Phi/kT.$$

With this equation, we can see the salient features of the Debye–Hückel theory beginning to emerge. First, the contribution of the ionic interactions to the energy will be negative, thus enhancing the stability of the solution. Second, the appropriate variable to describe ionic solutions is the ionic strength, $I = \frac{1}{2}\Sigma c_i z_i^2$, where c_i is a measure of the concentration. To render these features explicit and quantitative, requires a further step — the introduction of the Poisson equation. This is a standard equation of electrostatics which relates the second derivative of the electrostatic potential to the local charge density. In one dimension, we can express this as $d^2\Phi/dx^2 = -\rho/4\pi\varepsilon_0\varepsilon_r$. Substituting the equation we have obtained for the charge density into the radial form of the Poisson equation, Debye and Hückel obtained an equation that could be solved to give the additional electrostatic potential due to the ion–ion interactions as

$$\Phi = -\frac{zeb}{4\pi\varepsilon_0\varepsilon_r},$$

where

$$b^2 = \left(\frac{e^2}{\varepsilon_0\varepsilon_r kT}\right)\sum n_i^0 z_i^2.$$

$1/b$ has the dimensions of length and is known as the *Debye length*. It provides an approximate measure of the depth of the ion atmosphere, where the concentration of ions surrounding each individual ion is perturbed by the electrostatic forces. The additional energy of a single ion is then

$$dE = \int_0^q \Phi dq = \int_0^q \left(-\frac{z_i eb}{4\pi\varepsilon_0\varepsilon_r}\right) dq,$$

where $q = z_i e$, giving $dE = -z_i^2 e^2 b/8\pi\varepsilon_0\varepsilon_r$.

Assuming all the deviations from Henry's law at the dilute concentrations under consideration are due to the energy arising from electrostatic interactions and, since $dG = RT\ln\gamma_i$, we obtain

$$\ln\gamma_i = -\frac{z_i^2 e^2 b}{8\pi\varepsilon_0\varepsilon_r kT}.$$

Substituting for b, we obtain, for the experimentally-observable mean activity coefficients (Section 10.7),

$$\log\gamma_\pm = -|z_+ z_-|AI^{1/2},$$

where z_+ and z_- are the ionic charges on the anion and cation, respectively, and I is termed the *ionic strength*, defined by $I = \frac{1}{2}\Sigma z_i^2 m_i$, where $m_i = (\text{molality/mol kg}^{-1})$. This equation is referred to as the *Debye–Hückel limiting law*. A depends on the properties of the solvent and the temperature. In aqueous solutions at 25° C, $A = 0.509$. It tells us that the logarithm of the activity coefficient becomes more negative as the concentration of the solution increases and that the activity coefficients are less than unity. This means that it requires a higher concentration of ions to reach a given chemical potential than would be the case if there were no interactions between the ions. This is sometimes expressed as 'ions like being with other ions'. They can surround themselves with ions of opposite charge to obtain an energetically-favourable environment. Thus, more of a sparingly-soluble salt will dissolve in a dilute electrolyte solution than will dissolve in pure water (assuming there are no common ions).

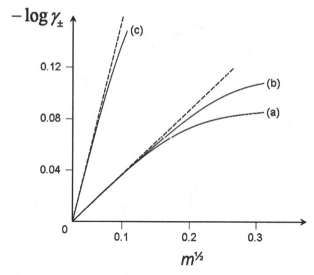

Fig. 10.13. Mean ionic activity coefficients as functions of the square root of molarity: (a) hydrogen chloride; (b) sodium chloride; (c) calcium chloride. The broken lines represent the Debye–Hückel limiting laws for $(z_+ = 2, z_- = 1)$ and $(z_+ = 1, z_- = 1)$ electrolytes.

In fact, the limiting equation only applies to very dilute solutions. (In fact, so dilute that they have been termed 'slightly contaminated distilled water'!) It can be extended by taking account of the fact that the ions are not point charges, by adding an additional term B which reflects the finite size of the ions,

$$\log \gamma_\pm = \frac{-A|z_+ z_-|I^{1/2}}{(1 + BI^{1/2})}.$$

With this modification, the *Debye–Hückel extended law*, the behaviour of ionic solutions up to about 0.1 mol dm^{-3} can be predicted — the actual range depends on the charges on the ions. In solutions containing more highly charged ions, the theory fails at lower concentrations (Fig. 10.13).

Ion atmospheres also affect the conductivity of ionic solutions by providing an electrostatic force which 'holds back' the moving ions and also by making them 'swim upstream' in the solvent which is being dragged in the opposite direction by ions of the opposite charge. A quantitative account of these effects which determine the conductivity of ionic solutions was given by Onsager in 1928.

10.9 Electrochemical cells

Reactions involving ions can be used to generate electricity in electrochemical cells. The change in Gibbs free energy accompanying the reaction, $\Delta_r G$, can be converted

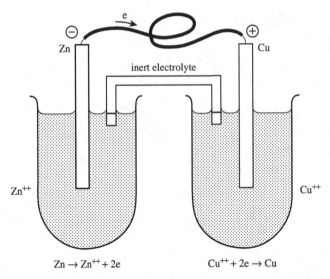

$$Zn \rightarrow Zn^{++} + 2e \qquad\qquad Cu^{++} + 2e \rightarrow Cu$$

Fig. 10.14. A simple electrochemical cell. The cell can be represented schematically, $Zn|ZnSO_4$ (aq)$\|CuSO_4$(aq)$|Cu$.

into electrical energy. If the cell is operated so the current flows infinitely slowly, i.e. under reversible conditions, we have

$$\Delta_r G = -nFE,$$

where F is the Faraday constant, $F = N_A e$, n is the number of moles of electrons transferred in the electrochemical reaction and E is the emf, the electrical potential generated by the cell.

Let us consider the cell in Fig. 10.14. At one electrode, electrons are liberated by the reaction and the electrode gains a negative charge. At this electrode, the process is one of *oxidation*, $Zn \rightarrow Zn^{2+} + 2e$. This electrode is termed the *anode*. At the other electrode, the *cathode*, electrons are consumed, $Cu^{2+} + 2e \rightarrow Cu$, and the process is *reduction*. The overall cell reaction is

$$Zn + Cu^{2+} \rightarrow Cu + Zn^{2+}.$$

In the *external* circuit, the electrons flow to the positive copper electrode from the negative zinc electrode. The cell can be represented by

$$Zn^{2+}(aq)/Zn\|Cu/Cu^{2+}(aq).$$

The symbol $\|$ represents the liquid junction, which it is assumed does not contribute to the electrode potential.

We can associate a *standard electrode potential* with each individual electrode reaction by defining each as the potential that would be measured against a standard

Table 10.1. Standard electrode potentials.

Electrode	Process	E/V
Li^+, Li	$Li^+ + e = Li$	-3.045
K^+, K	$K^+ + e = K$	-2.925
Na^+, Na	$Na^+ + e = Na$	-2.714
Mg^{++}, Mg	$Mg^{++} + 2e = Mg$	-2.37
Zn^{++}, Zn	$Zn^{++} + 2e = Zn$	-0.763
Fe^{++}, Fe	$Fe^{++} + 2e = Fe$	-0.44
Sn^{++}, Sn	$Sn^{4+} + 2e = Sn^{++}$	-0.136
Pb^{++}, Pb	$Pb^{++} + 2e = Pb$	-0.126
Fe^{3+}, Fe	$Fe^{3+} + 3e = Fe$	-0.036
H^+, H_2	$2H^+ + 2e = H_2$	[0]
Sn^{4+}, Sn^{++}	$Sn^{4+} + 2e = Sn^{++}$	0.15
Cu^{++}, Cu^+	$Cu^{++} + e = Cu+$	0.153
Cu^{++}, Cu	$Cu^{++} + 2e = Cu$	0.34
I_2, I^-	$I_2 + 2e = 2I^-$	0.536
Fe^{3+}, Fe^{2+}	$Fe^{3+} + e = Fe^{2+}$	0.771
Ag^+, Ag	$Ag^+ + e = Ag$	0.799
Hg^{2+}, Hg	$Hg^{2+} + 2e = Hg$	0.854
Br_2, Br^-	$Br_2 + 2e = 2Br^-$	1.0652
O_2, OH^-	$O_2 + 4H^+ + 4e = 2H_2O$	1.229
Cl_2, CI^-	$Cl_2 + 2e = 2CI^-$	1.3595

hydrogen electrode with all reactants in standard conditions of unit activity. The standard hydrogen electrode is formed by bubbling H_2 at 1 bar over a platinum electrode in the presence of H^+ ions at unit activity (Table 10.1).

Any positive standard electrode potential is a measure of the tendency of an element to gain electrons relative to the same tendency in hydrogen. If, for an electrode reaction $M \rightarrow M^+ + e$, the standard electrode potential is positive, then, since $\Delta_r G^\ominus = -nFE^\ominus$, the free energy will be negative and the reaction will tend to proceed as written. The electrode will lose electrons and become positively charged. For example, $E^\ominus(Cu^{2+}/Cu) = +0.34\,V$ and the process $Cu^{2+} + 2e \rightarrow Cu$ is favoured (relative to the standard hydrogen electrode). Conversely, a negative standard electrode potential such as that of zinc, $E^\ominus(Zn/Zn^{2+}) = -0.76\,V$, indicates that the zinc electrode will tend to gain electrons, $Zn^{2+} + 2e \rightarrow Zn$, and become negatively charged. Alkali metals have the most negative standard electrode potentials, whereas halogens have high positive values (Table 10.1). The standard cell potential of an electrochemical cell can be calculated by the difference between the standard electrode potentials of the two electrodes,

$$E^{\ominus} = E^{\ominus}_{Cu} - E^{\ominus}_{Zn} = 0.34 - (-0.76) = 1.10\,V.$$

$E^{\ominus}$ *is the standard electrical potential of the cell.* It is the electrical potential that would be generated when no current flows and when both reactants and products are present at unit activity.

We can write the expression for the free energy change accompanying a cell reaction in terms of the activities of the reactants and products,

$$\Delta_r G = \Delta_r G^{\ominus} + RT \ln \left(a_N^n a_M^m / a_A^a a_B^b \right) \quad \text{and, since} \quad \Delta_r G = -nFE,$$

we have

$$E = E^{\ominus} - \frac{RT}{nF} \ln \left(\frac{a_N^n a_M^m}{a_A^a a_B^b} \right).$$

This relationship is termed the *Nernst equation.* At equilibrium, the Gibbs free energy change for the reaction $\Delta_r G = 0$ and, consequently, $E = 0$ and

$$E^{\ominus} = \left(\frac{RT}{nF} \right) \ln K_{\ominus}.$$

$K_{\ominus}$ is the equilibrium constant for the cell reaction expressed in terms of activities.

When an electrochemical cell is not at equilibrium, a spontaneous reaction will occur and the cell will produce an electrical current flowing between the electrodes. However, when an external potential difference is applied to the electrodes, the process can be reversed and *electrolysis* will occur.

10.10 Summary of key principles

The concept of the *ideal solution* provides the basis for our treatment of the properties of real solutions and mixtures. An ideal solution is one which obeys *Raoult's law* so that the vapour pressure is a linear function of the concentrations of the components. In such a solution, the chemical potential of the components is given by $\mu_i = \mu_i^0 + RT \ln x_i$, where μ_i^0 is the chemical potential of pure liquid i. The molecular condition for ideality is that the two components of the solution are essentially identical.

True ideality is rare. However, in many dilute solutions, the solvent obeys Raoult's law, but the solute obeys *Henry's law*, so that its vapour pressure is linear but extrapolates to an arbitrary standard state which is not that of the pure liquid,

$$\mu_i = \mu_i^{\ominus} + RT \ln x_i.$$

We can understand this behaviour if we consider the molecular environment in a dilute solution. The solvent is surrounded largely by other solvent molecules whereas the solute is also surrounded by the solvent, rather than other solute molecules.

Real solutions often adhere to neither of these rules, but we retain the standard equations by replacing the true concentrations by *effective concentrations*, the activities, a_i, defined by $a_i = \gamma_i x_i$, where γ_i is the activity coefficient of component i. Attempts have been made to model the behaviour of real solutions by defining what are called *regular solutions* in which the entropy of mixing is ideal but the enthalpy of mixing is nonzero.

The behaviour of ions in solution is determined by the relative permittivity (dielectric constant) of the solvent, which reduces the energy of interaction of the ions, allowing ions of opposite charge to separate. Water has a very high value of relative permittivity, which facilitates this dissociation. The behaviour of ions in dilute solution was explained by Debye and Hückel in terms of an *ion atmosphere* of oppositely-charged ions that surrounds any given ion. They showed that this led to activity coefficients less than unity, indicating that the presence of other ions stabilised ions in solution. *'Ions like being with other ions'*. They derived the expression $\log \gamma_\pm = -|z_+ z_-| A I^{1/2}$.

Problems

(1) Calculate the ideal solubility of benzoic acid at 298 K. It melts at 395 K and has an enthalpy of fusion of 17.3 kJ mol^{-1}. Its solubility in benzene, measured as a mole fraction, is 0.082 and its solubility in ethanol is 0.19. Calculate its activity coefficient in benzene and comment on the solubility observed in ethanol.

(2) At 388 K, the vapour pressure of CCl$_4$ is 1.46 atm and that of SnCl$_4$ is 0.47 atm. Calculate the total vapour pressure and the mole fraction of carbon tetrachloride in the vapour phase of an equimolar liquid mixture. Assume that the mixture is ideal.

(3) Calculate the osmotic pressure of a 0.01 molar solution of sucrose at 298 K.

(4) Assuming that the Debye–Hückel limiting law applies, calculate the mean activity coefficients in 0.01 molar and 0.1 molar solutions of sodium chloride. The experimentally-observed values are 0.90 and 0.77.

(5) At 298 K, the standard potential of the electrochemical cell

$$Zn^{2+}(aq)/Zn \, \| \, Cu/Cu^{2+}(aq)$$

is 1.10 V. Calculate the standard free energy change accompanying the reaction

$$Cu^{2+}(aq) + Zn \rightarrow Cu + Zn^{2+}(aq).$$

11

Rates of Chemical Reactions

Thermodynamics enables us to determine the position at which chemical reactions will come to equilibrium. However, it gives us no information about how quickly or slowly that position of equilibrium will be reached. Indeed, some mixtures, far from equilibrium, will not react at any perceptible rate in the absence of a catalyst. In general, the rate of a chemical reaction is not directly dependent on the free energy difference between the reactants and products and is usually much slower than would be expected from the rates of encounter of the reactants. Most theories of reaction rates postulate an intermediate state of higher energy and many assume that this state is in thermodynamic equilibrium with the reactants.

11.1 The order of reactions

The first step in seeking a theoretical understanding of reaction rates is to eliminate the dependence on concentrations and to find an expression for the rate that is dependent only on temperature. The reaction rate is defined as the change in concentration of one of the reactants or products per unit time. In a reaction

$$aA + bB \rightarrow cC + dD,$$

the rate may be defined as

$$-\frac{d[A]}{dt}, \quad -\frac{d[B]}{dt}, \quad \frac{d[C]}{dt} \quad \text{or} \quad \frac{d[D]}{dt},$$

where the square brackets indicate concentrations. Frequently, the rate will depend in a simple way on the concentrations of the reactants but not necessarily in a manner related to the stoichiometry of the reaction.

Thus,

$$-\frac{d[A]}{dt} = k_r[A]^n[B]^m$$

would be a typical expression of the rate of the reaction illustrated above. n and m are often simple integers (or, sometimes, half integers) and k_r is a constant at any given temperature termed the *rate constant*.[1] The sum of the exponents, $(n+m)$, is defined as the *overall order of the reaction*. n and m are the orders with respect to each of the reactants and a knowledge of these orders, together with a knowledge of the rate constant, k_r, enables us to calculate the rate of the reaction when different concentrations of the reactants are present. A simple example is

$$H_2 + I_2 \rightarrow 2HI,$$

for which we find

$$\frac{d[HI]}{dt} = k_r[H_2][I_2].$$

This corresponds to an overall order of 2 and is first-order with respect to both H_2 and I_2. By contrast, for $H_2 + Br_2 \rightarrow 2HBr$, the reaction rate shows a complex dependence on H_2 and Br_2, even though the stoichiometric equation is the same as that for the formation of HI (Section 11.10). Another example is the reaction $S_2O_8^{2-} + 2I^- \rightarrow I_2 + 2SO_4^{2-}$, which might be expected to be a third-order reaction, but experiment shows it to be second-order.

Reactions are sometimes defined in terms of *molecularity*. This is a theoretical concept which indicates the number of molecules participating in an elementary step in a reaction. The molecularity and the order of a reaction are often different.

11.2 First-order reactions

The simplest type of kinetic behaviour is that represented by first-order rate equations. An example is radioactive decay of an unstable isotope, $X \rightarrow Y + \alpha$ or β radiation. The carbon ^{14}C nucleus decays in this way, with a *half-life*, the time for the initial concentration to be reduced to half its value, of some 5200 years.

An example of a chemical reaction which occurs with first-order kinetics is the decomposition of N_2O_5,

$$N_2O_5(g) \rightarrow 2NO_2(g) + \frac{1}{2}O_2(g).$$

---

[1] In this expression the rate constant is often written simply as k. We add the subscript r to avoid confusion with the Boltzmann constant.

For a first-order reaction of this type, the rate of reaction will be proportional to the quantity of reactant remaining and will follow the equation

$$\frac{d[X]}{dt} = -k_r[X], \quad \text{where [X] is the concentration of reactant at any time } t.$$

The rate constant, k_r, has the dimensions of reciprocal time and units of s^{-1}. The equation written as $d[X]/[X] = -k_r dt$ can be integrated to give

$$\ln[X] = -k_r t + c.$$

If we put $[X] = [X]_0$ at $t = 0$, the constant is given by $c = \ln[X]_0$ and

$$\ln\left\{\frac{[X]}{[X]_0}\right\} = -k_r t \quad \text{or}$$

$$[X] = [X]_0 \exp(-k_r t).$$

A linear plot of $\ln[X]$ or $\log[X]$ against t confirms the process is first-order. If $\ln[X]$ is plotted against t, the slope is equal to k_r, the rate constant as shown in Fig. 11.1. (For a plot of $\log[X]$ versus t, the slope is $k_r/2.303$.) Figure 11.2 shows that the time for half of X to disappear, the half-life, $t_{1/2}$, is constant. Thus, the time from $[X] = [X]_0/2$ to $[X] = [X]_0/4$ is also $t_{1/2}$. If we substitute $[X]/[X]_0 = 1/2$ into the expression for the rate, we obtain

$$t_{1/2} = \ln 2/k_r \approx 0.7/k_r.$$

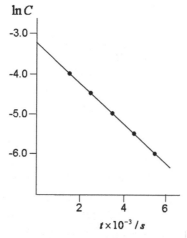

Fig. 11.1. A plot of log concentration versus time for the gaseous reaction $N_2O_5 \rightarrow 2NO_2 + \frac{1}{2}O_2$. The linear plot shows that the reaction is first order and the slope gives a rate constant of $5.0 \times 10^{-4}\,s^{-1}$.

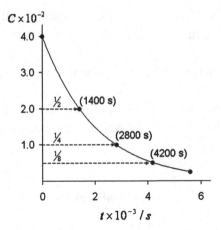

Fig. 11.2. A plot of concentration versus time for the reaction illustrated in Figure 11.1. Each of the half-lives is equal to (1400 s) consistent with first order kinetics.

11.3 Second-order reactions

The simplest second-order reactions are dimerisations, such as that of butadiene, which has been much studied:

$$2C_4H_6(g) \rightarrow C_8H_{12}(g).$$

This type of reaction can be represented, generally, by

$$A + A \rightarrow A_2,$$

for which the rate law is given by

$$-\frac{d[A]}{dt} = k_r[A]^2.$$

Integrating, we obtain

$$\frac{1}{[A]} = k_r t + \frac{1}{[A]_0},$$

where $[A]_0$ is the initial concentration of A. The second-order rate constant can be obtained by plotting $1/[A]$ against t to obtain a linear relationship of slope k_r. k_r has the dimensions of (concentration)$^{-1}$ (time)$^{-1}$ (Fig. 11.3). The rate law for second-order reactions shows that $t_{1/2}$ is not constant (as it is for first-order reactions), but is a function of [A], so that $t_{1/2} = 1/k_r[A]$ and, as [A] decreases, the half-life becomes longer.

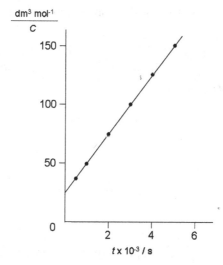

Fig. 11.3. A plot of the reciprocal of the concentration as a function of time for the dimerisation of butadiene. The linear plot shows that the reaction is second order and the slope is equal to the rate constant of the reaction.

We can represent simple second-order reactions, involving more than one reactant, such as $H_2 + I_2 \rightarrow 2HI$, by

$$A + B \rightarrow AB,$$

where

$$-\frac{d[A]}{dt} = k_r[A][B].$$

Since

$$d[A] = d[B], \quad [A]_0 - [A] = [B]_0 - [B].$$

If we simplify the expression by writing x as the decrease in A or B and writing a and b for the initial concentrations, that is $a = [A]_0$ and $b = [B]_0$, we obtain

$$\frac{dx}{dt} = k_r(a - x)(b - x),$$

where $(a - x)$ and $(b - x)$ are the concentrations of A and B at time t. On integration,

$$\left\{ \frac{1}{b - a} \right\} \ln \left\{ \frac{a(b - x)}{b(a - x)} \right\} = k_r t.$$

11.4 Determination of reaction order

There is no universally-applicable way to determine the order of a reaction and hence the rate equation that links the concentrations of reactants and products to time.

The most general method is to plot the concentrations as a function of time and to compare the result with the integrated form of the rate equations given in the previous two sections to find which fits the data best. This is the preferred method when the concentrations of several reactants appear in the rate equation. However, more direct methods are possible.

Half-lives: We have seen that $t_{1/2}$ is independent of concentration for a first-order reaction and inversely proportional to initial concentration, a, for a second-order reaction. In general,

$$\log t_{1/2} = -(n-1)\log a + \text{constant}$$

and a plot of $\log t_{1/2}$ versus a should be linear with a slope of $(1-n)$. This method is particularly valuable for non-integral orders. It can also be generalised to use fractions of the extent of a reaction other than one half. It is not applicable for the first-order reactions when $n = 1$.

Initial rates: If we can measure the rate of reaction, dx/dt, as a function of the initial concentration of the reactants, we can apply the equation

$$\frac{dx}{dt} = k_r[A]^n[B]^m$$

to determine the order with respect to each component. If we hold one component (say B) in excess, then

$$\frac{dx}{dt} = k_r'[A]^n$$

and the order, n, with respect to A can be determined from a plot of $\log(dx/dt)$ versus $\log[A]$. Then, by making A the reactant in excess, we can determine m, the order with respect to component B.

11.5 Effect of temperature on reaction rates

The rate constants of reactions are usually strongly dependent on temperature. Most typically, the rate increases with temperature according to a formula first proposed by Svante Arrhenius,

$$\frac{d \ln k_r}{dT} = \frac{E_a}{RT^2},$$

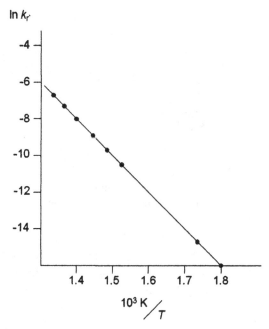

Fig. 11.4. A plot of log k_r against $1/T$ for the thermal decomposition of hydrogen iodide. The slope, $-E_a/R$, gives an activation energy of $184\,kJ\,mol^{-1}$.

where E_a is termed the *activation energy*. Integration leads to

$$\ln k_r = -\frac{E_a}{RT} + \text{constant}, \quad \text{giving}$$

$$k_r = Z\exp\left(\frac{-E_a}{RT}\right).$$

Z is referred to as the pre-exponential factor, which is often assumed to be independent of temperature. If the temperature dependence of a reaction follows the Arrhenius equation, a plot of $\ln k_r$ against $1/T$ is found to be linear with a slope of $-E_a/R$ (Fig. 11.4). The Arrhenius equation can be understood if we consider that only molecules with an energy above a critical amount, E_a, will be capable of reacting. A number of theories of reaction rates have been proposed to provide a quantitative interpretation of the pre-exponential factor.

11.6 Collision theory

The collision theory of gaseous chemical reactions considers the reacting molecules A and B as hard spheres of diameter σ_A and σ_B. If a molecule A moves

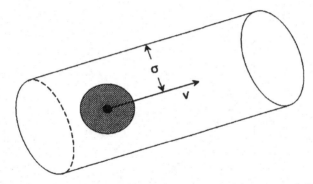

Fig. 11.5. Cylindrical space swept out by a moving molecule in one second. The cylinder is of length v, the molecular speed, and radius σ. Other molecules whose centres lie within the cylinder would undergo collision with the molecule.

with a mean relative speed v relative to a molecule B, it will sweep out a volume $\pi\sigma_{AB}^2 v$, where $\sigma_{AB} = (\sigma_A + \sigma_B)/2$ (Fig. 11.5). The number of collisions of an A molecule with B molecules, if the number of B molecules per unit volume is n_B, will be

$$\pi\sigma_{AB}^2 \, v \, n_B.$$

By arguments based on the kinetic theory of gases (Section 2.2), we can show that

$$v = \left[\left(\frac{8kT}{\pi}\right)\left(\frac{m_A + m_B}{m_A m_B}\right)\right]^{1/2},$$

where k is the Boltzmann constant. Under normal conditions of temperature and pressure, a molecule will collide on the order of 10^{10} times per second. If we multiply this result by n_A, the number of molecules of A per unit volume, we obtain the number of collisions of unlike molecules per second in a unit volume,

$$Z_{AB} = \left[\frac{8\pi kT(m_A + m_B)}{(m_A m_B)}\right]^{1/2} \sigma_{AB}^2 n_A n_B.$$

If the molecules of a single species react to produce products, then the colliding molecules are identical and we obtain (having introduced an additional factor of 1/2 to prevent counting every collision twice)

$$Z_{AA} = 2\left(\frac{\pi kT}{m}\right)^{1/2} \sigma^2 n^2.$$

At one bar pressure and near room temperature, $Z_{AA} \approx 10^{34} \, \text{m}^{-3} \, \text{s}^{-1}$. The rate of reaction, if every collision results in reaction, is given by Rate $= Z_{AA}$

$= k_r[A]^2 = k_r n^2$, where [A] is the concentration of the reactant and the corresponding (bi-molecular) rate constant is

$$k_r = 2 \left(\frac{\pi RT}{M} \right)^{1/2} \sigma^2, \quad \text{where } M \text{ is the molar mass.}$$

For unlike molecules reacting,

$$k_r = \left[\frac{8\pi RT(M_A + M_B)}{(M_A M_B)} \right]^{1/2} \sigma_{AB}^2.$$

Application of this equation to experimental data raises difficulties.

As an example, let us apply this equation to the decomposition of gaseous hydrogen iodide at 700 K and one bar pressure.

We estimate $\sigma = 400$ pm and the molar mass as 128×10^{-3} kg mol^{-1}.

$$k_r = 2 \left(\frac{\pi RT}{M} \right)^{1/2} \sigma^2 = 2 \left(\frac{3.14 \times 8.31 \times 700}{128 \times 10^{-3}} \right)^{1/2} (4 \times 10^{-10})^2 \, \text{m}^3 \, \text{s}^{-1}$$

$$k_r = 1.2 \times 10^{-16} \, \text{m}^3 \, \text{s}^{-1}.$$

Converting this to the units in which experimental data is more frequently expressed (moles of reaction with volumes measured in litres or dm^3), we obtain

$$k_r = 1.2 \times 10^{-16} \times 6.02 \times 10^{23} \times 10^3 \, \text{dm}^3 \, \text{s}^{-1} \, \text{mol}^{-1} = 7.3 \times 10^{10} \, \text{dm}^3 \, \text{s}^{-1} \, \text{mol}^{-1}.$$

The *observed rate is slower by a factor of 10^{14}* than these calculations, based only on the collision rate, would suggest. The agreement can be improved by the assumption, introduced in the previous section, that only molecular collisions at relative velocities corresponding to a kinetic energy greater than or equal to a critical energy E can react (Fig. 11.6). The fraction of such collisions is $\exp(-E/kT)$, leading to the

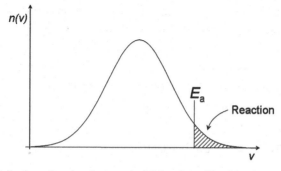

Fig. 11.6. The distribution of molecular speeds. Molecules with a kinetic energy greater than the activation energy, E_a, will react.

expression for the rate of reaction obtained by Arrhenius:

$$k_r = Z\,e^{-E/kT} \quad \text{and}$$

$$k_r = 2(\pi RT/M)^{1/2}\sigma^2\,e^{-E/kT}.$$

The activation energy, E, for a reaction is determined from the variation of the rate constant with temperature. We note that the $T^{1/2}$ in the expression for Z means that the experimental activation energy obtained from a plot of $\log k_r$ against $1/T$, E_a, is not quite the same energy, E, that occurs in the exponential term. In fact, $E_a = E - RT/2$. For a bi-molecular process, this typically leads to errors of $\sim 1\,\text{kJ}\,\text{mol}^{-1}$, a discrepancy which, in the context of experimental reaction rate studies, can often be ignored. In using this equation, we have assumed implicitly that the fact that the faster-moving molecules react does not significantly perturb the velocity distribution of the remaining reactant molecules. This assumption can be shown to be adequate if the energy, E, is greater than about five times the thermal energy, kT.

Returning to the decomposition of hydrogen iodide, the observed activation energy is $184\,\text{kJ}\,\text{mol}^{-1}$. Substituting this value in the term $e^{-E/kT}$, we find that

$$\exp[-184000/(8.31 \times 700)] = 1.83 \times 10^{-14},$$

leading to a predicted rate constant

$$k_r = 7.3 \times 10^{10} \times 1.83 \times 10^{-14} = 1.3 \times 10^{-3}\,\text{dm}^3\,\text{s}^{-1}\,\text{mol}^{-1}.$$

The experimentally-observed value is $1.6 \times 10^{-3}\,\text{dm}^3\,\text{s}^{-1}\,\text{mol}^{-1}$. The agreement is excellent, but unfortunately it is not typical. For most reactions, other than those involving the smallest atoms and molecules, the rates calculated in this manner are considerably greater than the values observed, and a further factor, p, the so-called *steric factor*, must be introduced. The pre-exponential factor becomes pZ. The steric factor was originally assumed to arise from the fact that the colliding molecules must be in a specific orientation for reaction to occur. However, since p cannot be evaluated from first principles, this does not provide a satisfactory theoretical interpretation. For the hydrogen iodide reaction, p was close to unity but it can frequently be as low as 10^{-5}, and occasionally much lower, thus a 'steric' interpretation is unconvincing. For example, in the reaction of hydrogen with ethene (ethylene) to produce ethane, the steric factor is found to be 1.7×10^{-6}.

11.7 Activated complex theory

The activated complex theory, often referred to as the *transition state theory*, has an advantage over collision theory in that it provides a more satisfactory interpretation

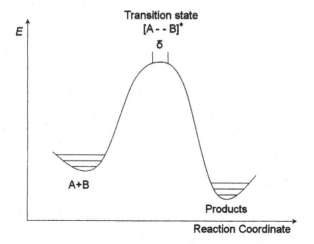

Fig. 11.7. The energy diagram for a reaction as described by the activated complex theory.

of the steric factor, p. It assumes that the reacting molecules come together to form an *activated complex*, which can be considered to be in equilibrium with the reacting molecules (Fig. 11.7). The rate of reaction, which is the rate of dissociation of the activated complex, is considered to be directly proportional to the concentration of the activated complex which, in turn, is determined by the equilibrium constant for the reaction leading to the formation of the activated complex.

$$A + B \rightleftharpoons [A \ldots B]^{\#}.$$

We can express the equilibrium constant in terms of the partition functions, z, of the reactants and the activated complex (see Section 8.10),

$$K = \frac{z_{A\ldots B}}{z_A z_B} \exp(-E/kT),$$

where E is the energy difference between the ground states of the activated complex and the reacting molecules. We can, to a reasonable approximation, equate this with the activation energy of the reaction, E_a. The partition functions of the reactants can be calculated in the normal way, but the partition function of the activated complex has a special feature. One vibration along the coordinate which corresponds to the dissociation of the activated complex is replaced in $z_{A\ldots B}$ by a translational contribution, $z^{\#}_{\text{trans}}$. The rate of dissociation of the activated complex can be written as $v^{\#}/\delta$, in terms of the velocity of translational motion along the reaction coordinate, $v^{\#}$, and an arbitrary distance, δ, which reflects the dimensions of the complex. (Fortunately, this arbitrary parameter cancels out so we do not need to pay much attention to its size.) Then, assuming half the activated complex dissociates to form reactants rather than products,

$$\text{Rate} = \frac{1}{2} \left(\frac{v^{\#}}{\delta} \right) K^{\#} z^{\#}_{\text{trans}} = \frac{1}{2} \left(\frac{v^{\#}}{\delta} \right) \frac{z^{\#} z^{\#}_{\text{trans}}}{z_A z_B} \exp \left(\frac{-E}{kT} \right).$$

$K^{\#}$ is defined by $K^{\#} = (z^{\#}/z_A z_B) \exp(-E/kT)$, where $z^{\#}$ is the partition function for the activated complex without the vibrational mode along the reaction coordinate. We can express the molecular partition functions in terms of the translational, vibrational and rotational components

$$z = z_{\text{trans}} z_{\text{rot}} z_{\text{vib}}, \quad \text{which we will write as } z = z_t z_r z_v.$$

We note that $v^{\#} = (2kT/\pi m^{\#})^{1/2}$, the average velocity in one dimension, and that the translational partition function is (Section 7.10)

$$z^{\#}_t = (2\pi m^{\#} kT)^{1/2} \left(\frac{\delta}{h} \right),$$

where δ is the space occupied by the activated complex (in one dimension). Thus,

$$k_r = \left(\frac{1}{2\delta} \right) \left(\frac{2kT}{\pi m^{\#}} \right)^{1/2} (2\pi m^{\#} kT)^{1/2} \left(\frac{\delta}{h} \right) K^{\#} \quad \text{and}$$

$$k_r = \left(\frac{kT}{h} \right) K^{\#}.$$

Our aim in applying transition state theory is to gain an understanding of the factors that determine reaction rates rather than a very precise evaluation of rate constants. We can simplify the evaluation of $K^{\#}$ by noting that the vibrational, rotational and translational partition functions are very different from each other in magnitude and each is reasonably similar from one molecule to another. To estimate $K^{\#}$, we need to consider the contributions to the partition functions from the various degrees of freedom possessed by the activated complex and the reaction molecules.

Let us consider the reaction of two atoms, noting that one vibrational degree of freedom is lost along the reaction coordinate.

$$A + B \rightarrow A \ldots B$$

Degrees of freedom: 3 trans + 3 trans → 3 trans + 2 rot (1 vib lost).

Therefore, expressing $K^{\#}$ in terms of the partition functions for each degree of freedom, we obtain

$$K^{\#} = \left(\frac{z_t^3 z_r^2}{z_t^3 z_t^3} \right) \exp \left(\frac{-E_a}{kT} \right) = \left(\frac{z_r^2}{z_t^3} \right) \exp \left(\frac{-E_a}{kT} \right).$$

The (one-dimensional) translational partition functions, z_t, depend on the size of the container and hence the concentration. From knowledge that $z_r \approx 10$ and $z_t \approx 10^{10}$, we can calculate the rate of the reaction as

$$k_r = \left(\frac{kT}{h}\right)\left(\frac{10^2}{10^{30}}\right)\exp\left(\frac{-E_a}{kT}\right)$$

and, since

$$\frac{kT}{h} \sim 10^{13}\,\text{s}^{-1},$$

we obtain

$$k_r = \left(\frac{10^{13} \times 10^2}{10^{30}}\right)\exp\left(\frac{-E_a}{kT}\right) = 10^{-15}\exp\left(\frac{-E_a}{kT}\right)\,\text{m}^3\,\text{s}^{-1}$$

which is of similar magnitude to that predicted by collision theory. In fact, if we substitute the expressions we determined earlier for z_r and z_t (Section 7.10), we find that the transition state theory applied to atomic reactions (where the reactants have no internal degrees of freedom) leads to an expression for k_r identical to that of the collision theory with the steric factor set at unity.

However, when we consider the reaction of molecules with internal degrees of freedom, very different results are obtained. Consider a reaction between two non-linear polyatomic molecules containing n atoms and m atoms, respectively, reacting to produce an activated complex with $(n + m)$ atoms.

	A	+	B	→	[AB]$^{\#}$
Degrees	3t		3t		3t
of	3r		3r		3r
Freedom	$(3n - 6)$ v		$(3m - 6)$ v		$(3n + 3m - 7)$ v

Again, we note that a vibrational degree of freedom along the reaction coordinate is lost.

$$k_r = \left(\frac{kT}{h}\right)\frac{(z_t^3 z_r^3 z_v^{(3n+3m-7)})}{(z_t^6 z_r^6 z_v^{(3n-6)} z_v^{(3m-6)})}\exp\left(-\frac{E_a}{kT}\right)$$

$$k_r = \frac{z_v^5}{z_t^3 z_r^3}\exp\left(-\frac{E_a}{kT}\right).$$

We can write this equation in the form

$$k_r = \left(\frac{z_v^5}{z_r^5}\right)\left(\frac{z_r^2}{z_t^3}\right)\exp\left(-\frac{E_a}{kT}\right).$$

As we have seen in the case of atomic reactions, if the pre-exponential factor is (z_r^2 / z_t^3), the steric factor is unity. We can therefore identify the steric factor, p, by

$$p = \left(\frac{z_v}{z_r} \right)^5$$

and, since $z_r \sim 10$ and $z_v \sim 1$, we obtain

$$p \sim 10^{-5}.$$

Values of this order of magnitude are often observed in the reactions of polyatomic molecules. For molecules of lesser complexity, such as linear diatomic molecules, intermediate values of the steric factor are predicted. The difficulty with the theory is that, whereas it is possible to investigate the structure and properties of the reactants so as to obtain the partition functions, it is difficult to obtain similar data for the activated complex. Therefore, it is not possible to use the theory to obtain an exact description of chemical reactions. Nevertheless, the activated complex theory enables us to have a good general understanding of the origins of the steric factor and the failings of simple collision theory.

Though we have assumed in deriving the theory that there is equilibrium between the activated complex and the reactants, this somewhat unrealistic assumption is not necessary. We need only make less stringent assumptions about the continuity of states between the reactants and the complex to obtain the same results.

11.8 Thermodynamic interpretation of activated complexes

We can treat the equilibrium between the activated complex and the reactants in a similar manner to that of any simple chemical equilibrium at constant pressure. We can write

$$-RT \ln K_P^{\#} = \Delta G^{\#},$$

where $\Delta G^{\#}$ is the Gibbs free energy of activation. The enthalpy of activation is then

$$\Delta H^{\#} = RT^2 \left(\frac{d \ln K_P^{\#}}{dT} \right).$$

For a gaseous reaction in which two reactants combine to form the activated complex (and where the change in the number of gaseous components is -1), we can write K in terms of concentrations, n_i / V (Section 8.7), such that

$$K_c^{\#} = K_P^{\#}(RT)$$

and, for one mole of reaction,

$$\Delta H^{\#} = RT^2 \left(\frac{\mathrm{d} \ln K_c^{\#}}{\mathrm{d}T} \right) - RT = E_a - RT$$

and

$$\Delta S^{\#} = \frac{\Delta H^{\#} - \Delta G^{\#}}{T}.$$

The rate constant can be expressed as

$$k_r = \left(\frac{kT}{h} \right) \exp \left(\frac{\Delta S^{\#}}{R} \right) \exp \left(-\frac{\Delta H^{\#}}{RT} \right).$$

If we ignore the difference between $\Delta H^{\#}$ and E_a, the steric factor can be interpreted in terms of the entropy of activation. The enthalpy and entropy of activation are the changes when the substances concerned are in the state defined by the unit of concentration employed for the rate constant. If we, rather arbitrarily, define the reaction rate in terms of $cm^3 \, mol^{-1}$, we obtain, conveniently, $\exp(\Delta S^{\#}/R) \approx p$. For $p = 1$, we calculate $\Delta S^{\#} \approx 0 \, J \, K^{-1} \, mol^{-1}$. This can be compared with a value observed for the simple bi-molecular decomposition of hydrogen iodide in the gas phase for which $\Delta S^{\#} = 0 \, J \, K^{-1} \, mol^{-1}$. For $p \sim 10^{-5}$, the value of the steric factor observed in many reactions between polyatomic molecules, we can estimate $\Delta S^{\#} \approx -100 \, J \, K^{-1} \, mol^{-1}$. This can be compared with the experimental value determined for the dimerisation of cyclopentadiene at 298 K, $\Delta S^{\#} = -110 \, J \, K^{-1} \, mol^{-1}$. (It should be noted that, had we chosen the more usual unit, $dm^3 \, mol^{-1}$, the entropies of activation would be some $57 \, J \, K^{-1} mol^{-1}$ more negative.)

11.9 Unimolecular reactions

A number of reactions, in particular some gas-phase decompositions, exhibit first-order kinetics — a feature not easily explained in terms of molecular collisions. Though the activation process must involve collisions, a bi-molecular process, the overall rate of reaction is observed to be independent of the collision rate. This paradox, which had perplexed chemists for many years, was resolved by the physicist F. A. Lindemann in the early 1920s. He considered the decomposition process taking place in two stages — activation by collision followed by a slow decomposition of the activated molecule.

$$A + A \rightarrow A^{\#} + A \quad k_1$$
$$A^{\#} + A \rightarrow A + A \quad k_{-1}$$
$$A^{\#} \rightarrow products \quad k_2$$

Lindemann proposed that most activated molecules would be deactivated by a further collision before they had time to dissociate. This being the case, we can make an approximation that is frequently very helpful in elucidating the rate laws of chemical reactions, called the *steady-state approximation*. This approximation makes the assumption that the concentrations of reaction intermediates, which are usually present only in very low concentrations, can be regarded as constant. In this case, if a steady state is reached, we make the assumption that the concentration of $[A^\#]$ is constant. Then,

$$\frac{d[A^\#]}{dt} = k_1[A]^2 - k_{-1}[A^\#][A] - k_2[A^\#] = 0,$$

$$[A^\#] = \frac{k_1[A]^2}{(k_{-1}[A] + k_2)}$$

and the rate of the reaction is given by

$$\text{Rate} = k_2[A^\#] = \frac{k_2 k_1[A]^2}{(k_{-1}[A] + k_2)}.$$

At high pressures, where [A] will be large, the rate of deactivation will be much greater than the rate of decomposition of the activated molecules, so that

$$k_{-1}[A^\#][A] \gg k_2[A^\#], \quad k_{-1}[A] \gg k_2,$$

the rate becomes

$$\text{Rate} = \frac{k_2 k_1[A]}{k_{-1}}$$

and *first-order kinetics* are observed. The Lindemann model provided a satisfactory explanation of how first-order kinetics can arise from bi-molecular collision processes.

At low pressures, when the time between collisions is long, most activated molecules will have time to decompose. Then

$$k_2[A^\#] \gg k_{-1}[A^\#][A], \quad k_2 \gg k_{-1}[A],$$

the rate is given by $\text{Rate} = k_1[A]^2$ and the reaction follows second-order kinetics.

It is possible to test this model in a simple way. We can write the general rate expression $\text{Rate} = k^*[A]$, where $k^* = k_2 k_1[A]/(k_{-1}[A] + k_2)$. This equation (the Lindemann equation) predicts that k^* should have a limiting value of $k_2 k_1/k_{-1}$ at high pressures, but will fall to zero at low pressures. Since [A] is proportional to the

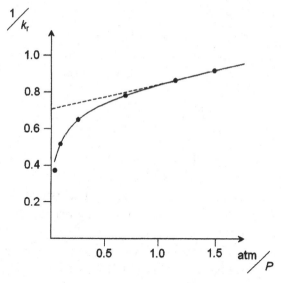

Fig. 11.8. Decomposition of azomethane at 603 K. The data (solid line) deviate from the Lindemann equation (broken line) at high pressures.

pressure P, we can write

$$\frac{1}{k^*} = \frac{k_{-1}}{k_1 k_2} + \frac{1}{k_1 P}$$

and a plot of $1/k^*$ versus $1/P$ should be linear.

At low pressures, the rate and order depend on pressure in the predicted manner (Fig. 11.8). However, at high pressures, the plot deviates from linearity and the reaction rates are often as much as 10^4 times faster than collision theory would predict. Cyril Hinshelwood proposed that the greater the number of vibrational modes present in the molecule, the greater was the chance of activation on collision. This was reflected in a higher value of k_1. Other improvements to the Lindemann theory sought to explain why the dissociation of the activated complex was found to be much faster than might be expected from the simple theory. One such improvement assumed that energy can flow freely from one vibrational mode to another throughout the activated molecules. The greater number of degrees of freedom the molecule possessed, the higher was the chance that it would reach the critical mode involved in dissociation, thus leading to a higher value of k_2 and to an increase in the rate of reaction. With these modifications, the theory gives an excellent account of the behaviour of unimolecular reactions.

11.10 Chain reactions

Rate laws do not always follow a simple order. Thus, whereas the reaction $H_2 + I_2 \rightarrow$ 2HI appears to follow simple second-order kinetics, the rate law for $H_2 + Br_2 \rightarrow$ 2HBr is far from simple and the rate law takes the form

$$\frac{d[HBr]}{dt} = \frac{k'[H_2][Br_2]^{1/2}}{(1 + k''[HBr]/[Br_2])}.$$

For the reaction $H_2 + Cl_2 \rightarrow$ 2HCl, no kinetic law can be established since, once initiated, the reaction proceeds with dramatic rapidity.

The clue to this complex behaviour can be understood by considering the hydrogen–bromine reaction. We will investigate the reaction in some detail, as it throws light on a number of important concepts that help our understanding of the rates of complex chemical reactions. If we investigate the *initial rate*, where no hydrogen bromide has yet been formed and $[HBr] = 0$, we find that

$$\frac{d[HBr]}{dt} = k[H_2][Br_2]^{1/2},$$

which would be the rate law for the elementary step

$$H_2 + Br \rightarrow HBr + H.$$

If bromine molecules and bromine atoms are in equilibrium, $Br_2 \rightleftharpoons 2Br$ and $[Br] \propto [Br_2]^{1/2}$.

This indicates that the reaction is initiated by bromine atoms. A subsequent reaction that occurs is

$$H + Br_2 \rightarrow HBr + Br.$$

The bromine atoms generated by this reaction can attack another hydrogen molecule and a *chain reaction* proceeds. However, when the reaction $H + HBr \rightarrow H_2 + Br$ occurs, the chain reaction does not proceed. We can write down all possible processes:

$$
\begin{array}{lll}
Br_2 \rightarrow 2Br & k_1 & \text{Chain initiation} \\
Br + H_2 \rightarrow HBr + H & k_2 & \text{Chain propagation} \\
H + Br_2 \rightarrow HBr + Br & k_3 & \text{Chain propagation} \\
H + HBr \rightarrow H_2 + Br & k_4 & \text{Chain inhibition} \\
2Br \rightarrow Br_2 & k_{-1} & \text{Chain termination}
\end{array}
$$

The rate of formation of hydrogen bromide can be written in terms of these reactions:

$$\frac{d[HBr]}{dt} = k_2[Br][H_2] + k_3[H][Br_2] - k_4[H][HBr].$$

The concentrations of [H] and [Br] are not easily measured and it is desirable to obtain an expression for the rate in terms of $[Br_2]$, $[H_2]$ and [HBr]. To do this, we can invoke the steady-state approximation, which makes the assumption that species in low concentration, such as [H] and [Br], can be assumed to be at a fixed concentration following a brief induction period.

(You may wish to ignore the algebra that follows and proceed directly to the result, the next highlighted equation below.)

If the concentration of hydrogen atoms is assumed to be constant,

$$\frac{d[H]}{dt} = k_2[Br][H_2] - k_3[H][Br_2] - k_4[H][HBr] = 0$$

giving

$$k_2[Br][H_2] = k_3[H][Br_2] + k_4[H][HBr]$$

and

$$[H] = \frac{k_2[H_2][Br]}{(k_3[Br_2] + k_4[HBr])}.$$

Substituting for $k_2[Br][H_2]$ in the equation for d[HBr]/dt, we obtain

$$\frac{d[HBr]}{dt} = 2k_3[H][Br_2].$$

If the concentration of bromine atoms is also assumed to be constant,

$$\frac{d[Br]}{dt} = 2k_1[Br_2] - 2k_{-1}[Br]^2 - k_2[Br][H_2]$$
$$+ k_3[H][Br_2] + k_4[H][HBr] = 0.$$

As

$$k_2[Br][H_2] = k_3[H][Br_2] + k_4[H][HBr]$$
$$\frac{d[Br]}{dt} = k_1[Br_2] - k_{-1}[Br]^2 = 0$$

and

$$[Br] = \left(\frac{k_1}{k_{-1}}\right)^{1/2} [Br_2]^{1/2}.$$

Substituting the values for [H] and [Br] in the rate equation for the formation of hydrogen bromide,

$$\frac{d[HBr]}{dt} = 2k_3[H][Br_2],$$

we obtain for the rate in terms of the concentrations of hydrogen, bromine and hydrogen bromide,

$$\frac{d[HBr]}{dt} = \frac{2k_2\,(k_1/k_{-1})^{1/2}\,[H_2][Br_2]^{1/2}}{(1 + k_4[HBr]/k_3[Br_2])}.$$

This is the rate expression that is observed experimentally.

Chain reactions are characterised by three steps:

(i) Creation of chain carriers, usually an atom or *radical*, R^*. (In the case of H_2 and Br_2 the chain carriers are Br and H atoms.)

(ii) Propagation of the chain

$$R^* + \text{reactants} \rightarrow \text{products} + R^*,$$

$$\text{e.g. } Br + H_2 \rightarrow HBr + H, \quad H + Br_2 \rightarrow HBr + Br$$

(iii) Termination of chain

$$R^* + R^* \rightarrow \text{nonreactive species}$$

$$\text{e.g. } Br + Br \rightarrow Br_2.$$

The effectiveness of the chain reaction is determined by the chain length — the number of propagation steps that can occur before the chain carrier is eliminated. In the case of $H_2 + Cl_2 \rightarrow 2HCl$, the chain length is longer than for the reaction of hydrogen with bromine, explaining the violence of the reaction.

The decomposition and oxidation of many organic molecules are chain reactions in which radicals are the chain carriers. For the decomposition of acetaldehyde (ethanal), the methyl radical is the chain carrier.

Initiation	$CH_3CHO \rightarrow CH_3^* + CHO^*$	k_1
Propagation	$CH_3CHO + CH_3^* \rightarrow CH_4 + CO + CH_3^*$	k_2
Termination	$2CH_3^* \rightarrow C_2H_6$	k_3

If we can regard the concentration $[CH_3^*]$ as constant and use the steady-state approximation, we obtain

$$\frac{d[CH_3^*]}{dt} = k_1[CH_3CHO] - k_3[CH_3^*]^2 = 0$$

giving

$$[CH_3^*] = \left(\frac{k_1}{k_3}\right)^{1/2} [CH_3CHO]^{1/2}$$

and

$$\frac{d[CH_4]}{dt} - k_2[CH_3CHO][CH_3^*]$$

$$= k_2 \left(\frac{k_1}{k_3}\right)^{1/2} [CH_3CHO]^{3/2}.$$

The predicted 3/2 order is in keeping with the experimental data. The existence of radicals in reactions of this type was originally confirmed by Friedrich Paneth, who showed that methyl radicals were released in the decomposition of lead tetramethyl, which, when flowing down a tube, could remove a deposit of lead. His pioneering observation was later confirmed by spectroscopy.

Chain reactions are of great practical importance in the production of polymers. Many common polymers, such as polyethylene and polystyrene, are formed by radical chain reactions. The process is initiated by the formation of a radical (R_1) which attaches to a double bond in the monomer, producing another radical which can itself react to continue the chain:

$$R_1 + M \rightarrow R_2 \quad \text{etc.}$$

$$R_n + M \rightarrow R_{(n+1)}$$

The process continues until two radicals combine to terminate the chain. The conditions under which the reaction is carried out determine the number of monomers that are linked and thus the properties of the polymer formed.

11.11 Explosions

Explosive reactions may occur in a number of ways. If an exothermic reaction occurs and heat is generated within the system, the increase in temperature will normally speed up the reaction, releasing even more heat. This may lead to a runaway reaction of ever-increasing violence which is called a *thermal explosion*. Frequently, explosions are caused by chain reactions in which the chain branches. That is, a reaction occurs in which one chain carrier can produce the product and more than one other chain carrier. The most famous example of a process of this type is not chemical, but is the nuclear bomb, where each uranium atom, when split, releases neutrons, which in turn cause further fission and liberate more neutrons.

A chemical example is the reaction

$$H_2 + O_2 \rightarrow 2H_2O.$$

The detailed mechanism of this reaction is very complicated, but it is possible to understand how explosion occurs in a fairly simple manner. Hydrogen atoms caused by the dissociation of the hydrogen molecule initiate two further reactions in which chain branching occurs as the number of chain carriers increases. Each radical generates two new radicals in the elementary steps

$$H + O_2 \rightarrow OH + O \quad \text{and}$$

$$O + H_2 \rightarrow OH + H.$$

If a stoichiometric mixture of oxygen and hydrogen is ignited, the conditions under which an explosion can occur depend on both the pressure and the temperature. At, for example, 770 K and at very low pressures, no explosion occurs, as the radicals can travel long distances and are eliminated before they can react on the walls of the container. At higher pressures, the mixture becomes explosive, but as the pressure is further increased a second limit is reached when gaseous collisions remove the chain carriers, and the reaction proceeds only slowly. Finally, at yet higher pressures, the heat generated by the reaction is unable to be dissipated causing the reaction to speed up, leading to a thermal explosion (Fig. 11.9).

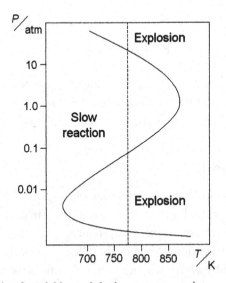

Fig. 11.9. Explosion limits of a stoichiometric hydrogen-oxygen mixture as a function of temperature and pressure. If the mixture at a temperature of 770 K is compressed (broken line), at approximately 0.002 atm it becomes explosive, then, above approximately 0.1 atm, the reaction proceeds only slowly. At considerably higher pressures, a third limit occurs and the mixture again becomes explosive.

11.12   Reactions in solution

Reactions in the gas phase have provided much of the theoretical understanding of reaction rates but, in the laboratory, reactions in solution are much more common. Reactions which take place in both the gaseous and liquid phases usually proceed at approximately the same pace. This is due to compensating factors. In the first place, in liquids, it takes time for reacting molecules to diffuse together, but, on the other hand, once in contact, they tend to undergo a number of collisions before separating, a phenomenon known as the *cage effect*. Molecules in a normal liquid may remain close to the same nearest neighbours for as long as 10^{-10} s. This effect is important for reactions with low activation energy such as the combination of radicals. If the rate of reaction is such that most of the reactant molecules have time to react before separating then the reaction rate is determined by the rate of diffusion in the solvent. Such reactions are termed *diffusion-controlled* reactions and rates will be directly proportional to the diffusion coefficients of the reactants which, in turn, will depend on the viscosity, η, of the solvent. The rate constants of diffusion-controlled reactions are given by

$$k_r \approx \frac{8000RT}{3\eta(T)}$$

and are typically on the order of 10^9 dm^3 mol^{-1} s^{-1} or greater. The activation energies for such reactions are typically about one third of the enthalpy of vaporization of the solvent. On the other hand, for reactions with high activation energy, the rate of reaction depends on the enthalpy of activation and the entropy of activation, which determine the rate of passage over the energy barrier. Such reactions are termed *activation-controlled reactions.*

Reactions involving ions are usually only observed in solution and the rates of reaction are very sensitive to the concentration of ions in solution and the charges on the ions. We can apply the transition state theory to substances in solution. In an ideal solution, we would express the equilibrium constant for the activated complex in terms of concentrations but, for real reactions, we need to formulate it in terms of activities,

$$K^{\#} = \frac{a^{\#}}{a_A a_B} = \left(\frac{c^{\#}}{c_A c_B}\right)\left(\frac{\gamma^{\#}}{\gamma_A \gamma_B}\right),$$

where a represents activities, c concentrations and γ the activity coefficients. We must be aware that an equation of this type requires us to define the standard state to which the activity coefficients refer. For most purposes, that will be a state behaving as if at infinite dilution but at a notional unit concentration (Section 10.7).

The activity coefficients depend on the ionic strength and, for ions in very dilute solutions, are predicted by the Debye–Hückel limiting law (Section 10.8). Consider two ionic species with charges z_A and z_B reacting via a transition state with charge $(z_A + z_B)$,

$$A^{z_A} + B^{z_B} \rightarrow [A \ldots B]^{z_A + z_B}.$$

The rate of reaction depends on the *concentration*, $c^{\#}$, giving

$$\text{Rate} = k_r c_A c_B = (kT/h)c^{\#} \quad \text{and} \quad k_r = (kT/h)K^{\#}(\gamma_A \gamma_B / \gamma^{\#}).$$

The reaction rate at infinite dilution, when the activity coefficients are unity, will be given by $k_r^0 = (kT/h)K^{\#}$ and, at finite concentrations, by

$$k_r = k_r^0 (\gamma_A \gamma_B / \gamma^{\#}).$$

Since, by the Debye–Hückel limiting law for aqueous solutions at 298 K, we have

$$\log \gamma = -0.509 z^2 \, I^{1/2} \quad \left(\text{where } I = \frac{1}{2} \Sigma(m_i z_i^2), \text{ the ionic strength} \right),$$

$$\log \left(\frac{k_r}{k_r^0} \right) = 0.509 \, I^{1/2} [(z_A + z_B)^2 - z_A^2 - z_B^2]$$

$$\log \left(\frac{k_r}{k_r^0} \right) = 2 \times 0.509 \, I^{1/2} z_A z_B = 1.018 \, I^{1/2} z_A z_B.$$

This equation predicts that when ions of the same charge react, the rate will increase as the ionic strength increases, whereas, when ions of opposite charge react, it will decrease. This variation of the rate with the ionic strength is called the *primary salt effect*. The agreement of the predictions of the Debye–Hückel theory with experimental data for low ionic strength is excellent (Fig. 11.10).

11.13 Catalysis

The strict definition of a catalyst can be very complicated, but for our purpose we can define a catalyst as a substance that speeds up a chemical reaction without altering the outcome or being used up in the process. Catalysts can reduce the activation energy of reactions very significantly and increase rates by many orders of magnitude.

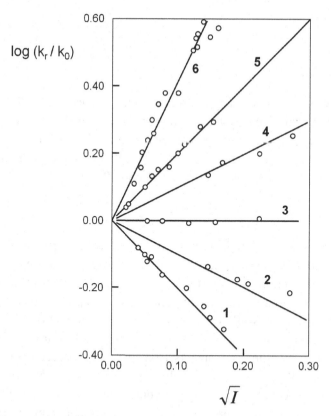

Fig. 11.10. The variation of the rates of ionic reactions with ionic strength. The circles are experimental values and the lines are the theoretical predictions given in Section 11.12. k_0 is the rate of the reactions at infinite dilution.

1: $[Co(NH_3)_5Br]^{2+} + OH^- \rightarrow [Co(NH_3)_5OH]^{2+} + Br^-$.
2: $H_2O_2 + 2H^+ + 2Br^- \rightarrow 2H_2O + Br_2$.
3: Inversion of sucrose.
4: $[NO_2NCOOC_2H_5]^- + OH^- \rightarrow C_2H_5OH + N_2O + CO_3^{2-}$.
5: $S_2O_8^{2-} + 2I^- \rightarrow I_2 + 2SO_4^{2-}$.
6: $2[Co(NH_3)_5Br]^{2+} + Hg^{2+} + 2H_2O \rightarrow 2[Co(NH_3)_5H_2O]^{3+} + HgBr_2$.

Acid-base catalysis: In solution, the most common form of catalysis is by acids and bases. Acid-base catalysed reactions are of considerable importance in organic and biological chemistry. In acid catalysis, a proton from the acid is transferred to the reactant, and catalysis by a base involves the transfer of a proton to the base. A common example is the hydrolysis of esters, which are catalysed by both acids and bases. The overall rate constant for the reaction

$$CH_3\,COOCH_3 + H_2O \rightarrow CH_3\,COOH + CH_3OH$$

can be written as

$$k_r = k_0 + k_H[\text{H}^+] + k_{\text{OH}}[\text{OH}^-],$$

where k_0 is the rate of the uncatalysed reaction.

Catalysis by solid surfaces: Many important catalytic reactions involve solid (often metallic or metal oxide) catalysts. One important mechanism is that of adsorption of the reactants producing a more favourable environment in which they can react. Langmuir gave a simple interpretation of the adsorption process to obtain the fraction of surface covered by gas molecules, θ, as (Section 9.13)

$$\theta = \frac{KP}{(1 + KP)}.$$

At low pressures, the coverage is directly proportional to pressure, but at high pressures the surface becomes saturated, and the coverage is independent of pressure. If we assume that reactions occur on the surface of a catalyst, then, for a unimolecular decomposition, we might expect that the rate would be proportional to θ, and, for a bi-molecular reaction, would be proportional to the product of the coverages of each of the reactants, i.e. $\theta_A \theta_B$. We can see that this can lead to quite complicated dependence on the pressures of reactants. Consider a simple decomposition, such as that of phosphine, PH_3, on tungsten. At low pressures, the surface is only sparsely covered so that the adsorption is directly proportional to pressure and the reaction is first-order. However, at higher pressures, the surface becomes almost fully saturated and the reaction is zero-order, that is, independent of the pressure. For bi-molecular reactions, even more possibilities exist. When molecules of two different species interact with a surface, one may restrict the adsorption of the other. Thus,

$$\theta_A = \frac{K_A P_A}{(1 + K_A P_A + K_B P_B)} \quad \text{and}$$

$$\theta_A \theta_B = \frac{K_A P_A K_B P_B}{(1 + K_A P_A + K_B P_B)^2}.$$

For example, the reaction of hydrogen and oxygen on platinum can be inhibited by high concentrations of oxygen which covers the surface and does not allow hydrogen to be adsorbed. Strongly adsorbed product molecules may also 'poison' the surface and hinder a reaction.

Enzyme catalysis: The catalysis of biological reactions by enzymes is kinetically very similar to reactions catalysed on surfaces. Again, the rate is found to vary linearly with reactant (substrate) concentration at low concentrations, but becomes independent of concentration at high concentrations. This was first explained by Michaelis and Menten. They assumed that the enzyme, E, and the substrate,

S, reacted to form a complex, ES, which subsequently decomposed, returning the enzyme and giving a product (P):

$$E + S \rightarrow ES \quad k_1$$
$$ES \rightarrow E + S \quad k_{-1}$$
$$ES \rightarrow E + P \quad k_2$$

Applying the steady-state approximation to the concentration of the enzyme-substrate complex leads to the expression

$$\frac{d[ES]}{dt} = k_1[E][S] - k_{-1}[ES] - k_2[ES] = 0.$$

The concentration of free enzyme is $[E] = [E]_0 - [ES]$, where $[E]_0$ is the initial concentration of enzyme. With enzyme reactions, the concentration of substrate, S, is usually much greater than that of enzyme and $[S] \gg [E]_0$, so we can assume that its concentration is not significantly reduced by binding to the enzyme. Eliminating $[E]$ between these two equations, we obtain

$$[ES] = \frac{k_1[E]_0[S]}{(k_{-1} + k_2 + k_1[S])}$$

and the rate is given by

$$Rate = k_2[ES] = \frac{k_2 k_1 [E]_0 [S]}{(k_{-1} + k_2 + k_1[S])},$$

which can be written

$$Rate = \frac{k_2[E]_0[S]}{(K_m + [S])}.$$

This expression is known as the *Michaelis–Menten equation* and K_m is the Michaelis constant, defined by

$$K_m = \frac{(k_{-1} + k_2)}{k_1}.$$

When $[S]$ is small and can be neglected compared with K_m, the kinetics become first-order with respect to the substrate concentration:

$$Rate = \frac{k_2[E]_0[S]}{K_m}.$$

When $[S]$ is much greater than K_m, the rate is independent of substrate concentration and is limited by the availability of the enzyme, so the kinetics become zero-order:

$$Rate = k_2[E]_0.$$

In general, a plot of the reciprocal of the rate against $1/[S]$, called a *Lineweaver–Burk plot*, is linear and enables K_m and $k_2[E]$ to be determined. The rate law is similar in form to that obtained from the Langmuir isotherm for gases reacting on surfaces.

11.14 Reaction dynamics

Most modern investigations into the mechanisms of chemical reactions have been concerned with a detailed analysis of reactive collisions and attempts to understand how the rate of reaction depends on the internal states of the individual reactants. Reactions can be studied in crossed molecular beams (Fig. 11.11), which enable the direction and the energy with which the reaction product emerges to be analysed. Such investigations show that reactive collisions exhibit a wide variety of behaviour.

The results for the reaction

$$H + Cl_2 \rightarrow HCl + CL$$

show that much of the HCl is scattered backwards in the direction of the incoming hydrogen beam (Fig. 11.12). Such reactions, which are dominated by the repulsive

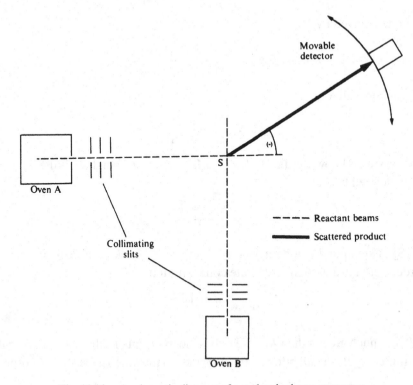

Fig. 11.11. A schematic diagram of a molecular beam apparatus.

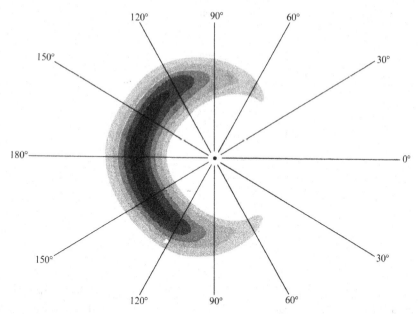

Fig. 11.12. A map of the density of HCl scattered following collisions of hydrogen atoms with chlorine molecules. The scattering is predominantly backward along the direction of the incoming hydrogen atom (180°).

forces between the reactants, are termed *rebound reactions*. Much of the energy of the reaction appears in the translational energy of the HCl. By contrast, in the reaction

$$K + I_2 \rightarrow KI + I,$$

much of the product is scattered in the forward direction in what is called a *stripping reaction* (Fig. 11.13). In this case, little of the energy goes into product recoil and most goes into internal modes, largely the vibrational excitation of the KI. The reaction has a very high cross-section, which means that reaction can occur even if the potassium atoms are on a trajectory that would take them some distance from the iodine molecules. This is explained by what is known as the *harpoon mechanism*. As the reactants approach, an electron is transferred — at a relatively large distance — from the potassium to the iodine molecule, and the oppositely-charged species are then pulled together to react. The long-range transfer of the electron, the harpoon, explains the high reaction cross-section.

In the reaction

$$O + Br_2 \rightarrow BrO + Br,$$

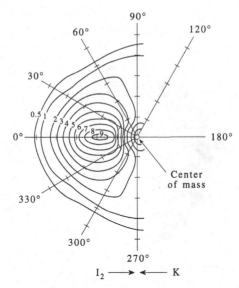

Fig. 11.13. A contour map of the velocity distribution for KI formed from collisions between potassium atoms and iodine molecules. In this "stripping reaction", the product molecules continue in the direction of the incident potassium atoms.

the product, BrO, does not show any tendency to be scattered in either the forward or backward direction. This implies that the reaction complex, $Br\cdots O\cdots Br$, must be sufficiently long-lived compared with its period of rotation so that any 'memory' of the incoming direction of the oxygen atoms is lost.

Lasers allow individual energy states of the reactants to be selected and the energy levels of the product analysed. In studying the reaction

$$HCl + K \rightarrow KCl + H,$$

the HCl laser can be used to excite the hydrogen chloride molecules into the first excited vibrational state. In this state, the reaction is found to proceed a hundred times faster than when the hydrogen chloride is in its ground vibrational state. It is also possible to examine the vibrational state distributions of the products of reactions in molecular beams by examining the infrared emission.

11.15 Photochemical reactions

When a molecule absorbs a photon, the excited state formed can behave in a number of ways (Section 6.16). It can lose the energy of excitation by, for example, fluorescence or phosphorescence, returning to its ground state by what is referred to as a photophysical process. On the other hand, some excited states can take part in chemical reactions. When reactions are initiated by the absorption of radiation, the products can be different from when the same reaction is excited thermally. The photochemical excitation commonly involves a narrow energy range and can lead to a wide array of thermodynamically unstable products.

The quantum yield, Φ, of a reaction

$$A + h\nu \rightarrow L + M$$

can be defined, for a specific product L, in terms of the number of molecules of that species generated from the absorption of one photon by molecule A,

$$\Phi_L = \frac{\text{(Number of molecules of L produced)}}{\text{(Number of quanta absorbed by A)}}$$

(with both quantities measured over the same volume and time). This gives

$$\Phi_L = \frac{d[L]/dt}{I},$$

where I is the intensity of the radiation absorbed.

For the photodecomposition of dimethylketone (acetone) in the vapour phase,

$$(CH_3)_2CO + h\nu \rightarrow C_2H_6 + CO \quad \text{(and other products)},$$

it is found that, for a wide range of incident wavelengths, pressure and temperature, $\Phi_{CO} = 1$. However, for other chemical reactions, the secondary processes which can follow the primary photophysical processes can lead to quantum yields far from unity. For

$$H_2 + Cl_2 \rightarrow 2HCl,$$

the chain reaction (Section 11.10) which follows the photodissociation of chlorine into radicals can lead to values of Φ_{HCl} on the order of 10^6. Conversely, if activated molecules lose the energy of excitation before reaction takes place, the quantum yield will be less than unity. The reactivity of excited states depends not only on their intrinsic reactivity, but also on their lifetime. States that are rapidly quenched do not have time to react. Photochemical reactions often involve the lowest triplet states of molecules, which have a relatively long lifetime.

A detailed study of photochemical reactions is beyond the scope of this text. Photochemical reactions play an important role in many aspects of biochemistry and chemistry. They are involved in atmospheric reactions, including the causes of atmospheric pollution. They have also influenced the evolution of life on Earth, in particular through the generation of ozone, which protects organisms from harmful UV radiation in the range 230–290 nm.

11.16 Summary of key principles

The rates of chemical reactions can normally be defined in terms of the concentrations of the components and a rate constant, k_r:

$$\text{Rate} = k_r[A]^n[B]^m.$$

n and m are termed the *orders of the reaction* with respect to components A and B, and $(n+m)$ is the overall order. The rates of chemical reactions are usually strongly dependent on temperature. If we assume that only molecules with greater than a certain critical energy react, we obtain the Arrhenius equations

$$\frac{d \ln k_r}{dT} = \frac{E_a}{RT^2}, \quad \ln k_r = -\frac{E_a}{RT} + \text{constant} \quad \text{and} \quad k_r = Z \exp\left(-\frac{E_a}{RT}\right),$$

where E_a is the *activation energy* of the reaction and Z is referred to as the pre-exponential factor.

Collision theory seeks to interpret the rate of chemical reactions in terms of the frequency of molecular collisions. In pure substances, under standard conditions, the collision rate is of the order of $10^{34}\,\text{m}^{-3}\text{s}^{-1}$, leading to reaction rates many orders of magnitude faster than those observed. Making allowance for the fact that only molecules with kinetic energy greater than the activation energy react removes much of the discrepancy. Collision theory then provides a reasonable account of reactions between atoms and very simple molecules. However, when more complicated molecules react, it can predict rates too fast by a factor of 10^5 or more. The discrepancy is identified as a so-called *steric factor*, which modifies the pre-exponential factor. *Activated complex theory* provides a method of estimating the steric factor by evaluating the equilibrium constant for the activated complex and the reactants. The steric factor can be expressed in terms of the rotational and vibrational partition functions.

Activated molecules may take time to react and, if this time is long compared with the time between collisions, they may be deactivated before reaction can take place. Under these circumstances, the reaction will appear first-order and a number of these so-called *unimolecular reactions* occur. As the pressure is reduced, a point is reached when all the activated molecules decompose and the reaction becomes second-order.

Some reactions do not follow simple mechanisms and their rate laws can be very complicated. Many of these are *chain reactions*, where a chain carrier, often a radical, is regenerated by the reaction. The rate of reaction can depend in a very complex way on the concentrations of reactants and products. The expressions for the rate are usually obtained by making use of the steady-state approximation, which assumes that reaction intermediates, such as radicals, are present in only small quantities and their concentration reaches a steady state.

Due to compensating factors, *reactions in solution* often proceed at very much the same rate as those in the gas phase. Reactions with low activation energy tend to be controlled by the speed of diffusion of the reactants in the solvent, whereas those with higher activation energy are limited by the fraction of molecules that have sufficient energy.

Catalysts, which increase the rate of a reaction without being consumed in the process, can act either in solution or on solid surfaces. Many biological processes are catalysed by enzymes.

Problems

(1) The initial rates for the reaction between CO and Cl_2 at 645 K were determined as follows:

P_{Cl_2}/atm	0.875	0.636	0.303	0.118	0.118	0.118
P_{CO}/atm	0.116	0.116	0.116	0.842	0.611	0.276
Rate $\times 10^3$/atm s^{-1}	1.89	1.16	0.38	0.68	0.50	0.22

Calculate the order of the reaction with respect to carbon monoxide and to chlorine.

(2) The conversion of ortho-hydrogen to para-hydrogen at 923 K is observed to have the following half lives:

Pressure/mm Hg	50	100	200	400
Half life/s	645	450	318	222

Determine the order of the reaction.

(3) The rate constants for a unimolecular decomposition reaction were as follows:

T/K	823	843	863	883
k/s^{-1}	4.25	13.94	39.3	98

Estimate the activation energy of the reaction.

(4) A reaction is 10% complete in 600 s at 310 K and 120 s at 330 K. Estimate the activation energy of the reaction.

(5) The following first-order rate constants were observed for the decomposition of $(CH_3)_2O$ at 777 K:

Pressure/mm Hg	28	58	91	150	171
k/s^{-1}		0.88	3.50	3.26	4.13	4.61

To what extent does the reaction satisfy the Lindemann equation?

(6) The rate constant for the reaction between X ions and Ag^+ ions in water at 298 K depends on the ionic strength of the solution, I, as:

$k \times 10^{10}/dm^3\,mol^{-1}\,s^{-1}$	3.50	3.30	3.05	2.85
$I/mol\,dm^{-3}$	10^{-4}	10^{-3}	5×10^{-3}	10^{-2}

Use these data to determine the charge on the ion X.

Answers to Problems

2.1 $310\,\mathrm{m\,s^{-1}}$

2.2 44

2.3 $-0.9\,\mathrm{kJ}$, $-5.7\,\mathrm{kJ}$

2.4 $46.0\,\mathrm{kJ\,mol^{-1}}$

2.5 $-84.4\,\mathrm{kJ\,mol^{-1}}$

2.6 $-1.15 \times 10^{-18}\,\mathrm{J}$, $-1.4 \times 10^{-20}\,\mathrm{J}$

3.1 $0.30\,\mathrm{MJ}$

3.2 $4 \times 10^{-12}\,\mathrm{m}$

3.3 $\sim0.5\,\mathrm{nm}$

3.4 $1.46 \times 10^{-21}\,\mathrm{J}$, $0.88\,\mathrm{kJ}$

3.5 $109700\,\mathrm{cm^{-1}}$, $27430\,\mathrm{cm^{-1}}$

4.1 2.85

4.2 $554\,\mathrm{kJ\,mol^{-1}}$

4.3 $216\,\mathrm{pm}$

5.1 Bond orders N_2, Cl_2, Cl_2^+ 3, 1, 1.5.

5.2 $2 \times 10^{-4}\,\mathrm{cm}$

5.3 $\alpha \pm \beta$

6.1 $1.098 \times 10^5\,\mathrm{cm^{-1}}$, $27\,450\,\mathrm{cm^{-1}}$

6.2 $120\,\mathrm{cm^{-1}}$

6.3 $12276\,\mathrm{cm^{-1}}$

6.4 $152\,\mathrm{pm}$

6.5 $6.428 \times 10^{13}\,\mathrm{sec^{-1}}$, $12.82\,\mathrm{kJ\,mol^{-1}}$

7.1 $-1.54\,\mathrm{J\,K^{-1}}$

7.2 $14.2\,\mathrm{J\,K^{-1}}$

7.3 $111\,\mathrm{J\,K^{-1}}$

7.4 $35\,\mathrm{J\,K^{-1}}$

7.5 1.07×10^{32}, $178\,\mathrm{kJ\,mol^{-1}}$

7.6 19.3

8.1 $-142\,\mathrm{kJ\,mol^{-1}}$, $-197\,\mathrm{kJ\,mol^{-1}}$, 7×10^{24}

8.2 83% isobutane

8.3 $12\,\mathrm{kJ\,mol^{-1}}$

8.4 $62\,\mathrm{kJ\,mol^{-1}}$, $0.19\,\mathrm{kJ\,K^{-1}\,mol^{-1}}$

8.5 $63\,\mathrm{kJ\,mol^{-1}}$, 4×10^{-5}

9.1 $35.5\,\mathrm{kJ\,mol^{-1}}$

9.2 $1.5\,\mathrm{K}$

9.3 $388\,\mathrm{pm}$

9.4 face-centred cubic

9.5 $447\,\mathrm{m^2}$

10.1 2.2

10.2 $0.97\,\mathrm{atm}$, 0.76

10.3 $0.25\,\mathrm{bar}$

10.4 0.89, 0.69

10.5 $-212\,\mathrm{kJ\,mol^{-1}}$

11.1 1, 1.5

11.2 1.5

11.3 $316\,\mathrm{kJ\,mol^{-1}}$

11.4 $68\,\mathrm{kJ\,mol^{-1}}$

11.5 $1/k$ vs $1/P$ is approximately linear

11.6 X^-, (slope $= -1$)

Appendix 1
Thermochemical Data at 298.15 K

The thermodynamic quantities listed are for one mole of substance in the standard state, that is, at one bar pressure. The enthalpies and Gibbs free energy of formation of the substances are the changes in these properties when a substance in its standard state is formed from its elements in their standard states. The standard state of an element is the normal physical state at one bar pressure and, for the data given in these tables, at 298.15 K. The entropies listed are absolute and are based on the assumption that the entropy of a pure substance is zero at the absolute zero of temperature.

Substance	$\Delta_f H^0/$ kJ mol^{-1}	$\Delta_f G^0/$ kJ mol^{-1}	$S^0/$ J K^{-1} mol^{-1}	$C_p^0/$ J K^{-1} mol^{-1}
AgCl(s)	−127.07	−109.77	96.2	50.79
Ar(g)	0.00	0.00	154.84	20.79
Br(g)	118.88	82.40	175.02	20.79
Br$_2$(g)	30.91	3.11	245.46	36.02
C(s) (graphite)	0.00	0.00	5.74	8.53
CCl$_4$(l)	−135.44	−65.2	216.40	131.75
CH$_4$(g)	−74.81	−50.72	186.26	35.31
CO(g)	−110.53	−137.17	197.67	29.14
CO$_2$(g)	−393.51	−394.36	213.74	37.11
C$_2$H$_2$(g)	226.73	209.20	200.94	43.93
C$_2$H$_4$(g)	52.26	68.15	219.56	43.56
C$_2$H$_6$(g)	−84.68	−32.82	229.60	52.63
n-C$_4$H$_{10}$(g)	−126.15	−17.03	310.23	97.45
i-C$_4$H$_{10}$(g)	−134.53	−20.92	294.6	96.82
CH$_3$OH(l)	−238.66	−166.27	126.8	81.6

(*Continued*)

(Continued)

Substance	$\Delta_f H^0/$ kJ mol^{-1}	$\Delta_f G^0/$ kJ mol^{-1}	$S^0/$ J K^{-1} mol^{-1}	$C_P^0/$ J K^{-1} mol^{-1}
CS$_2$(l)	89.70	65.27	151.34	75.7
C$_2$H$_5$OH(l)	−227.69	−174.78	160.7	111.46
c-C$_3$H$_6$(g)	53.30	104.45	237.55	55.94
n-C$_3$H$_6$(g)	−103.85	−23.49	269.91	73.5
CH$_3$COOH(l)	−484.5	−389.9	159.8	124.3
C$_6$H$_6$(l)	49.0	124.3	173.3	136.1
H(g)	217.97	203.25	114.71	20.78
H$_2$(g)	0.00	0.00	130.68	28.82
HBr(g)	−36.40	53.45	198.70	29.14
HCl(g)	−92.31	−95.30	186.91	29.12
HI(g)	26.48	1.70	206.59	29.16
H$_2$O(l)	−285.83	−237.13	69.91	75.29
H$_2$O(g)	−241.82	−228.57	188.83	33.58
H$_2$S(g)	−20.63	−33.56	205.79	33.23
Kr(g)	0.00	0.00	164.08	20.79
I (g)	106.84	70.25	180.79	20.79
I$_2$(g)	62.44	19.33	260.69	36.90
N(g)	472.70	455.56	153.30	20.79
N$_2$(g)	0.00	0.00	191.61	29.13
NH$_3$(g)	−46.11	−16.45	192.45	35.06
NO(g)	90.25	86.55	210.79	29.84
NO$_2$(g)	33.18	51.31	240.06	37.20
N$_2$O(g)	82.05	104.20	219.85	38.45
N$_2$O$_4$(g)	9.16	97.89	304.29	77.28
Na(s)	0.00	0.00	51.21	28.24
NaCl(s)	−411.15	−384.14	72.13	50.50
NaOH(s)	−425.61	−379.49	64.46	59.54
O(g)	249.17	231.73	160.06	21.91
O$_2$(g)	0.00	0.00	205.14	29.36
S(s) rhombic	0.00	0.00	31.80	22.64
SO$_2$(g)	−296.83	−300.19	248.22	39.87
SO$_3$(g)	−395.72	−371.06	256.76	50.67
Xe(g)	0.00	0.00	169.68	20.79

Appendix 2
Hydrogen-Like Wave Functions

$n = 1, \quad l = 0, \quad m_l = 0$

$$1s = \left(\frac{1}{\pi}\right)^{\frac{1}{2}} \left(\frac{Z}{a_0}\right)^{\frac{3}{2}} \exp\left(\frac{-Zr}{a_0}\right)$$

$n = 2, \quad l = 0, \quad m_l = 0$

$$2s = \left(\frac{1}{32\pi}\right)^{\frac{1}{2}} \left(\frac{Z}{a_0}\right)^{\frac{3}{2}} \left(2 - \frac{Zr}{a_0}\right) \exp\left(\frac{-Zr}{2a_0}\right)$$

$n = 2, \quad l = 1, \quad m_l = 0$

$$2p_z = \left(\frac{1}{32\pi}\right)^{\frac{1}{2}} \left(\frac{Z}{a_0}\right)^{\frac{3}{2}} \left(\frac{Zr}{a_0}\right) \exp\left(\frac{-Zr}{2a_0}\right) \cos\theta$$

$n = 2, \quad l = 1, \quad m_l = \pm 1$

$$2p_{x,y} = \left(\frac{1}{64\pi}\right)^{\frac{1}{2}} \left(\frac{Z}{a_0}\right)^{\frac{3}{2}} \left(\frac{Zr}{a_0}\right) \exp\left(\frac{-Zr}{2a_0}\right) \exp(\pm i\varphi)$$

$n = 3, \quad l = 0, \quad m_l = 0$

$$3s = \left(\frac{1}{81}\right) \left(\frac{1}{3\pi}\right)^{\frac{1}{2}} \left(\frac{Z}{a_0}\right)^{\frac{3}{2}} \left[27 - 18\left(\frac{Zr}{a_0}\right) + 2\left(\frac{Zr}{a_0}\right)^2\right] \exp\left(\frac{-Zr}{3a_0}\right)$$

$n = 3, \quad l = 1, \quad m_l = 0$

$$3p_z = \left(\frac{1}{81}\right) \left(\frac{2}{\pi}\right)^{\frac{1}{2}} \left(\frac{Z}{a_0}\right)^{\frac{3}{2}} \left(\frac{Zr}{a_0}\right) \left[6 - \left(\frac{Zr}{a_0}\right)\right] \exp\left(\frac{-Zr}{3a_0}\right) \cos\theta$$

$n = 3, \quad l = 1, \quad m_l = \pm 1$

$$3p_{x,y} = \left(\frac{1}{81}\right) \left(\frac{2}{\pi}\right)^{\frac{1}{2}} \left(\frac{Z}{a_0}\right)^{\frac{3}{2}} \left(\frac{Zr}{a_0}\right) \left[6 - \left(\frac{Zr}{a_0}\right)\right] \exp\left(\frac{-Zr}{3a_0}\right) \sin\theta \exp(\pm i\varphi)$$

$n = 3, \quad l = 2, \quad m_l = 0$

$$3d_z = \left(\frac{1}{81}\right)\left(\frac{1}{6\pi}\right)^{\frac{1}{2}}\left(\frac{Z}{a_0}\right)^{\frac{3}{2}}\left(\frac{Zr}{a_0}\right)^2 \exp\left(\frac{-Zr}{3a_0}\right)(3\cos^2\theta - 1)$$

$$n = 3, \quad l = 2, \quad m_l = \pm 1$$

$$3d_{x,y} = \left(\frac{1}{81}\right)\left(\frac{2}{\pi}\right)^{\frac{1}{2}}\left(\frac{Z}{a_0}\right)^{\frac{3}{2}}\left(\frac{Zr}{a_0}\right)^2 \exp\left(\frac{-Zr}{3a_0}\right)\sin\theta\cos\theta\exp(-i\varphi)$$

$$n = 3, \quad l = 2, \quad m_l = \pm 2$$

$$3d_{x,y} = \left(\frac{1}{81}\right)\left(\frac{1}{2\pi}\right)^{\frac{1}{2}}\left(\frac{Z}{a_0}\right)^{\frac{3}{2}}\left(\frac{Zr}{a_0}\right)^2 \exp\left(\frac{-Zr}{3a_0}\right)\sin^2\theta\exp(-i\varphi)$$

Appendix 3
Symmetry

Symmetry can provide much information about molecular properties and often provides a powerful insight into many of the problems that arise in physical chemistry. The study of symmetry in advanced physical chemistry is called group theory and is beyond the scope of this book. We will confine ourselves to outlining a few simple symmetry elements that can be used to characterise molecules and the arrangements of electrons within them. Each symmetry element is associated with a symmetry operation, that is, a movement which, when performed, will restore an object to its original configuration.

(i) *Inversion.* When each part of a molecule, with coordinates (x, y, z), can be transported to points with coordinates $(-x, -y, -z)$ and this operation leaves the molecule unchanged, the molecule is said to possess a *centre of symmetry*. Molecules such as CO_2 and SF_6 possess a centre of symmetry.

(ii) *Reflection.* Another symmetry element is when reflection in a *mirror plane* leaves the molecule unchanged. H_2O does not possess a centre of symmetry, but has two mirror planes through which reflection leaves the molecule unchanged.

(iii) *Rotation.* A molecule may possess an *axis of symmetry* so that, when rotated about this axis, its original configuration is restored. We express the degree of rotation required to return the molecule to its original state in terms of n, where a rotation of $360°/n$ produces the same configuration. A molecule such as CH_3Cl possesses a three-fold axis of rotation about the carbon-chlorine bond. If a molecule possesses more than one axis of rotation, the one with the largest value of n is said to be the principal axis.

There are five symmetry elements in total, but the three described above are the only ones referred to in the text. Molecules can be classified according to their symmetry elements. The set of symmetry elements possessed by a molecule constitute a *point group*. The point group to which a molecule belongs determines many of its properties.

Appendix 4
Units and Fundamental Constants

Length	ångstrom, Å	10^{-10} m
	micron, μ	10^{-6} m
Volume	litre, l, L	1 dm^3
Pressure	bar	10^5 Pa
	atmosphere, atm	1.013×10^5 Pa
	mm Hg, torr	1.333×10^2 Pa (1 atm/760)
Energy*	electronvolt, eV	1.602×10^{-19} J
		(96.485 kJ mol^{-1})
	hartree, E_h	4.360×10^{-18} J
		(2.625×10^3 kJ mol^{-1})

*See also the energy conversion table.

Energy conversion table

	joule	kJ mol^{-1}	cm^{-1}
1 joule	1	6.022×10^{20}	5.034×10^{22}
1 kJ mol^{-1}	1.660×10^{-21}	1	83.594
1 cm^{-1}	1.986×10^{-23}	1.196×10^{-2}	1

Prefixes

tera, T 10^{12}	deci, d 10^{-1}	nano, n 10^{-9}
giga, G 10^9	centi, c 10^{-2}	pico, p 10^{-12}
mega, M 10^6	milli, m 10^{-3}	femto, f 10^{-15}
kilo, k 10^3	micro, μ 10^{-6}	
deca, da 10^1		

SI units

Length	l	metre, m
Mass	m	kilogram, kg
Time	t	second, s
Electric current	I	ampere, A
Temperature	T	kelvin, K
Force	F	newton, N, kg m s^{-2}
Pressure	P	pascal, Pa, kg m^{-1} s^{-2} (N m^{-2})
Energy	E, U	joule, J, kg m^2 s^{-2}
Charge	q	coulomb, C, As
Magnetic field strength	B	tesla, T, kg s^{-2} A^{-1}

Approximate values of physical constants

Atomic mass constant	m_u, (amu)	1.661×10^{-27} kg
Avogadro constant	N_A	6.022×10^{23} mol^{-1}
Bohr magneton	μ_B	9.274×10^{-24} J T^{-1}
Bohr radius	a_0	5.292×10^{-11} m
Boltzmann constant	k	1.381×10^{-23} J K^{-1}
Charge on proton	e	1.602×10^{-19} C
Faraday constant	F	9.649×10^4 C mol^{-1}
Gas constant	R	8.314 J K^{-1} mol^{-1}
Nuclear magneton	μ_N	5.051×10^{-27} J T^{-1}
Vacuum permittivity	ε_0	8.854×10^{-12} J^{-1} C^2 m^{-1}
	$4\pi\varepsilon_0$	1.113×10^{-10} J^{-1} C^2 m^{-1}
Planck constant	h	6.626×10^{-34} J s
Rydberg constant	R_H	1.09737×10^5 cm^{-1}
Mass of electron	m_e	9.109×10^{-31} kg
Mass of proton	m_p	1.673×10^{-27} kg
Speed of light (vacuum)	c	2.998×10^8 m s^{-1}

Data for 298.15 K

Molar volume (perfect gas at 1 bar) $= 24.79$ dm^3

$RT = 2.479$ kJ mol^{-1}

$RT/F = 25.69$ mV

$kT/hc = 207.23$ cm^{-1}

Moments of inertia: conversion factor

$$I/\text{kg m}^2 = 1.661 \times 10^{-47}(I/\text{amu Å}^2)$$

Further Reading

The texts identified immediately below provide additional material and fuller accounts of the subject matter of the chapter to which they refer. They have been selected because they are of approximately the same standard of difficulty as this text and are not necessarily chosen to take the subject to a higher level. That aspect is dealt with below in the section entitled *The way forward*.

Thermodynamics: Chapters 2, 7, 8, 10

E. B. Smith, *Basic Chemical Thermodynamics*, 5th Edition (2004). Imperial College Press, London.

J. M. Seddon and J. D. Gale, *Thermodynamics and Statistical Mechanics* (2001). Royal Society of Chemistry, London.

R. P. H. Gasser and W. G. Richards, *An Introduction to Statistical Thermodynamics* (1995). Imperial College Press, London.

Quantum mechanics: Chapter 3

D. O. Hayward, *Quantum Mechanics for Chemists* (2002). Royal Society of Chemistry, London.

Atomic and molecular structure: Chapters 4 and 5

W. G. Richards and P. R. Scott, *Structure and Spectra of Atoms* (1976). Wiley, London.

R. McWeeny, *Coulson's Valence* (1979). Oxford University Press, Oxford.

J. Barrett, *Structure and Bonding* (2001). Royal Society of Chemistry, London.

Spectra: Chapter 6

C. N. Banwell and E. M. McCash, *Fundamentals of Molecular Spectra* (1994). McGraw Hill, New York. [Despite its title, this book covers both atomic and molecular spectra.]

Photochemistry: Chapters 6 and 11

R. P. Wayne, *Principles and Applications of Photochemistry* (1988). Oxford University Press, Oxford.

States of matter: Chapter 9

M. Rigby, E. B. Smith, W. A. Wakeham and G. C. Maitland, *The Forces Between the Molecules* (1986). Clarendon Press, Oxford.

P. J. Wheatley, *The Determination of Molecular Structure* (1981). Dover Publications, New York. [This book gives a very useful account of diffraction methods.]

Electrochemistry: Chapters 9 and 11

D. R. Crow, *Principles and Applications of Electrochemistry* (1994). Blackie, Glasgow. [Includes electrokinetic processes.]

Reaction Kinetics: Chapter 11

M. J. Pilling and P. W. Seakins, *Reaction Kinetics* (1995). Oxford University Press, Oxford.

J. I. Steinfield, J. S. Francisco and W. L. Hase, *Chemical Kinetics and Dynamics* (1998). Prentice Hall, New Jersey.

M. R. Wright, *Chemical kinetics* (2004). Wiley, Chichester.

The way forward

For almost three quarters of a century, physical chemistry has benefited from excellent English language textbooks. After the Second World War, that of Samuel Glasstone was the definitive text. This was replaced in the 1960s by that of Walter J. Moore. These books still retain some interest for connoisseurs of the subject, but, at the present time, probably the most influential textbook is that of Peter Atkins. This provides a comprehensive account of physical chemistry and anyone familiar with its contents could regard themselves as a professional physical chemist. This present book is intended to provide a first step on the road on which the major textbooks are the next staging post. In addition to the Atkins textbook, I would recommend, for theoretical aspects of the subject, the book by Donald McQuarrie and John Simon.

P. Atkins and J. de Paula, *Atkins' Physical Chemistry,* 9th Edition (2009). Oxford University Press, Oxford.

D. A. McQuarrie and J. D. Simon, *Physical Chemistry: A Molecular Approach* (1997). University Science Books, Sausalito, California.

IA	IIA	IIIA	IVA	VA	VIA	VIIA	VIII			IB	IIB	IIIB	IVB	VB	VIB	VIIB	0
$_{1}$H 1·008																	$_{2}$He 4·003
$_{3}$Li 6·941	$_{4}$Be 9·012											$_{5}$B 10·81	$_{6}$C 12·01	$_{7}$N 14·01	$_{8}$O 16·00	$_{9}$F 19·00	$_{10}$Ne 20·18
$_{11}$Na 22·99	$_{12}$Mg 24·31											$_{13}$Al 26·98	$_{14}$Si 28·09	$_{15}$P 30·97	$_{16}$S 32·06	$_{17}$Cl 35·45	$_{18}$Ar 39·95
$_{19}$K 39·10	$_{20}$Ca 40·08	$_{21}$Sc 44·96	$_{22}$Ti 47·90	$_{23}$V 50·94	$_{24}$Cr 52·00	$_{25}$Mn 54·94	$_{26}$Fe 55·85	$_{27}$Co 58·93	$_{28}$Ni 58·71	$_{29}$Cu 63·55	$_{30}$Zn 65·37	$_{31}$Ga 69·72	$_{32}$Ge 72·59	$_{33}$As 74·92	$_{34}$Se 78·96	$_{35}$Br 79·90	$_{36}$Kr 83·80
$_{37}$Rb 85·47	$_{38}$Sr 87·62	$_{39}$Y 88·91	$_{40}$Zr 91·22	$_{41}$Nb 92·91	$_{42}$Mo 95·94	$_{43}$Te 98·91	$_{44}$Ru 101·1	$_{45}$Rh 102·9	$_{46}$Pd 106·4	$_{47}$Ag 107·9	$_{48}$Cd 112·4	$_{49}$In 114·8	$_{50}$Sn 118·7	$_{51}$Sb 121·8	$_{52}$Te 127·6	$_{53}$I 126·9	$_{54}$Xe 131·3
$_{55}$Cs 132·9	$_{56}$Ba 137·3	$_{57}$La 138·9	$_{72}$Hf 178·5	$_{73}$Ta 180·9	$_{74}$W 183·9	$_{75}$Re 186·2	$_{76}$Os 190·2	$_{77}$Ir 192·2	$_{78}$Pt 195·1	$_{79}$Au 197·0	$_{80}$Hg 200·6	$_{81}$Tl 204·4	$_{82}$Pb 207·2	$_{83}$Bi 209·0	$_{84}$Po (210)	$_{85}$At (210)	$_{86}$Rn (222)
$_{87}$Fr (223)	$_{88}$Ra 226·0	$_{89}$Ac (227)															

Lanthanides	$_{57}$La 138·9	$_{58}$Ce 140·1	$_{59}$Pr 140·9	$_{60}$Nd 144·2	$_{61}$Pm (147)	$_{62}$Sm 150·4	$_{63}$Eu 152·0	$_{64}$Gd 157·3	$_{65}$Tb 158·9	$_{66}$Dy 162·5	$_{67}$Ho 164·9	$_{68}$Er 167·3	$_{69}$Tm 168·9	$_{70}$Yb 173·0	$_{71}$Lu 175·0
Actinides	$_{89}$Ac (227)	$_{90}$Th 232·00	$_{91}$Pa 231·0	$_{92}$U 238·0	$_{93}$Np 237·0	$_{94}$Pu (242)	$_{95}$Am (243)	$_{96}$Cm (248)	$_{97}$Bk (247)	$_{98}$Cf (251)	$_{99}$Es (254)	$_{100}$Fm (253)	$_{101}$Md (256)	$_{102}$No (254)	$_{103}$Lw (257)

Index